These safety symbols are used in laboratory and field investigations in this book to indicate poss...  **W9-CFA-144**
ing of each symbol and refer to this page often. *Remember to wash your hands thoroughly after completing lab procedures.*

## PROTECTIVE EQUIPMENT   Do not begin any lab without the proper protection equipment.

 **GOGGLES** Proper eye protection must be worn when performing or observing science activities that involve items or conditions as listed below.

**APRON** Wear an approved apron when using substances that could stain, wet, or destroy cloth.

 **SOAP** Wash hands with soap and water before removing goggles and after all lab activities.

 **GLOVES** Wear gloves when working with biological materials, chemicals, animals, or materials that can stain or irritate hands.

## LABORATORY HAZARDS

| Symbols | Potential Hazards | Precaution | Response |
|---|---|---|---|
| **DISPOSAL** | contamination of classroom or environment due to improper disposal of materials such as chemicals and live specimens | • DO NOT dispose of hazardous materials in the sink or trash can.<br>• Dispose of wastes as directed by your teacher. | • If hazardous materials are disposed of improperly, notify your teacher immediately. |
| **EXTREME TEMPERATURE** | skin burns due to extremely hot or cold materials such as hot glass, liquids, or metals; liquid nitrogen; dry ice | • Use proper protective equipment, such as hot mitts and/or tongs, when handling objects with extreme temperatures. | • If injury occurs, notify your teacher immediately. |
| **SHARP OBJECTS** | punctures or cuts from sharp objects such as razor blades, pins, scalpels, and broken glass | • Handle glassware carefully to avoid breakage.<br>• Walk with sharp objects pointed downward, away from you and others. | • If broken glass or injury occurs, notify your teacher immediately. |
| **ELECTRICAL** | electric shock or skin burn due to improper grounding, short circuits, liquid spills, or exposed wires | • Check condition of wires and apparatus for fraying or uninsulated wires, and broken or cracked equipment.<br>• Use only GFCI-protected outlets | • DO NOT attempt to fix electrical problems. Notify your teacher immediately. |
| **CHEMICAL** | skin irritation or burns, breathing difficulty, and/or poisoning due to touching, swallowing, or inhalation of chemicals such as acids, bases, bleach, metal compounds, iodine, poinsettias, pollen, ammonia, acetone, nail polish remover, heated chemicals, mothballs, and any other chemicals labeled or known to be dangerous | • Wear proper protective equipment such as goggles, apron, and gloves when using chemicals.<br>• Ensure proper room ventilation or use a fume hood when using materials that produce fumes.<br>• NEVER smell fumes directly.<br>• NEVER taste or eat any material in the laboratory. | • If contact occurs, immediately flush affected area with water and notify your teacher.<br>• If a spill occurs, leave the area immediately and notify your teacher. |
| **FLAMMABLE** | unexpected fire due to liquids or gases that ignite easily such as rubbing alcohol | • Avoid open flames, sparks, or heat when flammable liquids are present. | • If a fire occurs, leave the area immediately and notify your teacher. |
| **OPEN FLAME** | burns or fire due to open flame from matches, Bunsen burners, or burning materials | • Tie back loose hair and clothing.<br>• Keep flame away from all materials.<br>• Follow teacher instructions when lighting and extinguishing flames.<br>• Use proper protection, such as hot mitts or tongs, when handling hot objects. | • If a fire occurs, leave the area immediately and notify your teacher. |
| **ANIMAL SAFETY** | injury to or from laboratory animals | • Wear proper protective equipment such as gloves, apron, and goggles when working with animals.<br>• Wash hands after handling animals. | • If injury occurs, notify your teacher immediately. |
| **BIOLOGICAL** | infection or adverse reaction due to contact with organisms such as bacteria, fungi, and biological materials such as blood, animal or plant materials | • Wear proper protective equipment such as gloves, goggles, and apron when working with biological materials.<br>• Avoid skin contact with an organism or any part of the organism.<br>• Wash hands after handling organisms. | • If contact occurs, wash the affected area and notify your teacher immediately. |
| **FUME** | breathing difficulties from inhalation of fumes from substances such as ammonia, acetone, nail polish remover, heated chemicals, and mothballs | • Wear goggles, apron, and gloves.<br>• Ensure proper room ventilation or use a fume hood when using substances that produce fumes.<br>• NEVER smell fumes directly. | • If a spill occurs, leave area and notify your teacher immediately. |
| **IRRITANT** | irritation of skin, mucous membranes, or respiratory tract due to materials such as acids, bases, bleach, pollen, mothballs, steel wool, and potassium permanganate | • Wear goggles, apron, and gloves.<br>• Wear a dust mask to protect against fine particles. | • If skin contact occurs, immediately flush the affected area with water and notify your teacher. |
| **RADIOACTIVE** | excessive exposure from alpha, beta, and gamma particles | • Remove gloves and wash hands with soap and water before removing remainder of protective equipment. | • If cracks or holes are found in the container, notify your teacher immediately. |

# Your online portal to everything you need

connectED.mcgraw-hill.com

Look for these icons to access exciting digital resources

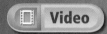 Video

 Audio

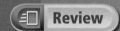 Review

? Inquiry

WebQuest

✓ Assessment

Concepts in Motion

ANIMALS

i SCIENCE

Glencoe

**ANIMALS**

**SCIENCE**

Glencoe

**Snow Leopard,** *Uncia uncia*

The snow leopard lives in central Asia at altitudes of 3,000 m–5,500 m. Its thick fur and broad, furry feet are two of its adaptations that make it well suited to a snowy environment. Snow leopards cannot roar but can hiss, growl, and make other sounds.

*The McGraw-Hill Companies*

 **Education**

Send all inquiries to:
McGraw-Hill Education
8787 Orion Place
Columbus, OH 43240-4027

ISBN: 978-0-07-888015-5
MHID: 0-07-888015-7

Printed in the United States of America.

3 4 5 6 7 8 9 10 11 12   DOW   15 14 13 12

# Authors and Contributors

## Authors

**American Museum of Natural History**
New York, NY

**Michelle Anderson, MS**
Lecturer
The Ohio State University
Columbus, OH

**Juli Berwald, PhD**
Science Writer
Austin, TX

**John F. Bolzan, PhD**
Science Writer
Columbus, OH

**Rachel Clark, MS**
Science Writer
Moscow, ID

**Patricia Craig, MS**
Science Writer
Bozeman, MT

**Randall Frost, PhD**
Science Writer
Pleasanton, CA

**Lisa S. Gardiner, PhD**
Science Writer
Denver, CO

**Jennifer Gonya, PhD**
The Ohio State University
Columbus, OH

**Mary Ann Grobbel, MD**
Science Writer
Grand Rapids, MI

**Whitney Crispen Hagins, MA, MAT**
Biology Teacher
Lexington High School
Lexington, MA

**Carole Holmberg, BS**
Planetarium Director
Calusa Nature Center and
Planetarium, Inc.
Fort Myers, FL

**Tina C. Hopper**
Science Writer
Rockwall, TX

**Jonathan D. W. Kahl, PhD**
Professor of Atmospheric Science
University of Wisconsin-
Milwaukee
Milwaukee, WI

**Nanette Kalis**
Science Writer
Athens, OH

**S. Page Keeley, MEd**
Maine Mathematics and
Science Alliance
Augusta, ME

**Cindy Klevickis, PhD**
Professor of Integrated Science
and Technology
James Madison University
Harrisonburg, VA

**Kimberly Fekany Lee, PhD**
Science Writer
La Grange, IL

**Michael Manga, PhD**
Professor
University of California, Berkeley
Berkeley, CA

**Devi Ried Mathieu**
Science Writer
Sebastopol, CA

**Elizabeth A. Nagy-Shadman, PhD**
Geology Professor
Pasadena City College
Pasadena, CA

**William D. Rogers, DA**
Professor of Biology
Ball State University
Muncie, IN

**Donna L. Ross, PhD**
Associate Professor
San Diego State University
San Diego, CA

**Marion B. Sewer, PhD**
Assistant Professor
School of Biology
Georgia Institute of Technology
Atlanta, GA

**Julia Meyer Sheets, PhD**
Lecturer
School of Earth Sciences
The Ohio State University
Columbus, OH

**Michael J. Singer, PhD**
Professor of Soil Science
Department of Land, Air and
Water Resources
University of California
Davis, CA

**Karen S. Sottosanti, MA**
Science Writer
Pickerington, Ohio

**Paul K. Strode, PhD**
I.B. Biology Teacher
Fairview High School
Boulder, CO

**Jan M. Vermilye, PhD**
Research Geologist
Seismo-Tectonic Reservoir
Monitoring (STRM)
Boulder, CO

**Judith A. Yero, MA**
Director
Teacher's Mind Resources
Hamilton, MT

**Dinah Zike, MEd**
Author, Consultant,
Inventor of Foldables
Dinah Zike Academy;
Dinah-Might Adventures, LP
San Antonio, TX

**Margaret Zorn, MS**
Science Writer
Yorktown, VA

## Consulting Authors

**Alton L. Biggs**
Biggs Educational Consulting
Commerce, TX

**Ralph M. Feather, Jr., PhD**
Assistant Professor
Department of Educational
Studies and Secondary
Education
Bloomsburg University
Bloomsburg, PA

**Douglas Fisher, PhD**
Professor of Teacher Education
San Diego State University
San Diego, CA

**Edward P. Ortleb**
Science/Safety Consultant
St. Louis, MO

## Series Consultants

### Science

**Solomon Bililign, PhD**
Professor
Department of Physics
North Carolina Agricultural
and Technical State University
Greensboro, NC

**John Choinski**
Professor
Department of Biology
University of Central Arkansas
Conway, AR

**Anastasia Chopelas, PhD**
Research Professor
Department of Earth and
Space Sciences
UCLA
Los Angeles, CA

**David T. Crowther, PhD**
Professor of Science Education
University of Nevada, Reno
Reno, NV

**A. John Gatz**
Professor of Zoology
Ohio Wesleyan University
Delaware, OH

**Sarah Gille, PhD**
Professor
University of California
San Diego
La Jolla, CA

**David G. Haase, PhD**
Professor of Physics
North Carolina State
University
Raleigh, NC

**Janet S. Herman, PhD**
Professor
Department of Environmental
Sciences
University of Virginia
Charlottesville, VA

**David T. Ho, PhD**
Associate Professor
Department of Oceanography
University of Hawaii
Honolulu, HI

**Ruth Howes, PhD**
Professor of Physics
Marquette University
Milwaukee, WI

**Jose Miguel Hurtado, Jr.,
PhD**
Associate Professor
Department of Geological
Sciences
University of Texas at El Paso
El Paso, TX

**Monika Kress, PhD**
Assistant Professor
San Jose State University
San Jose, CA

**Mark E. Lee, PhD**
Associate Chair & Assistant
Professor
Department of Biology
Spelman College
Atlanta, GA

**Linda Lundgren**
Science writer
Lakewood, CO

v

## Series Reviewers, continued

**Carolyn Elliott**
Iredell-Statesville Schools
Statesville, NC

**Christine M. Jacobs**
Ranger Middle School
Murphy, NC

**Jason O. L. Johnson**
Thurmont Middle School
Thurmont, MD

**Felecia Joiner**
Stony Point Ninth Grade Center
Round Rock, TX

**Joseph L. Kowalski, MS**
Lamar Academy
McAllen, TX

**Brian McClain**
Amos P. Godby High School
Tallahassee, FL

**Von W. Mosser**
Thurmont Middle School
Thurmont, MD

**Ashlea Peterson**
Heritage Intermediate Grade Center
Coweta, OK

**Nicole Lenihan Rhoades**
Walkersville Middle School
Walkersvillle, MD

**Maria A. Rozenberg**
Indian Ridge Middle School
Davie, FL

**Barb Seymour**
Westridge Middle School
Overland Park, KS

**Ginger Shirley**
Our Lady of Providence Junior-Senior High School
Clarksville, IN

**Curtis Smith**
Elmwood Middle School
Rogers, AR

**Sheila Smith**
Jackson Public School
Jackson, MS

**Sabra Soileau**
Moss Bluff Middle School
Lake Charles, LA

**Tony Spoores**
Switzerland County Middle School
Vevay, IN

**Nancy A. Stearns**
Switzerland County Middle School
Vevay, IN

**Kari Vogel**
Princeton Middle School
Princeton, MN

**Alison Welch**
Wm. D. Slider Middle School
El Paso, TX

**Linda Workman**
Parkway Northeast Middle School
Creve Coeur, MO

## Teacher Advisory Board

The Teacher Advisory Board gave the authors, editorial staff, and design team feedback on the content and design of the Student Edition. They provided valuable input in the development of *Glencoe ⓘScience.*

**Frances J. Baldridge**
Department Chair
Ferguson Middle School
Beavercreek, OH

**Jane E. M. Buckingham**
Teacher
Crispus Attucks Medical
Magnet High School
Indianapolis, IN

**Elizabeth Falls**
Teacher
Blalack Middle School
Carrollton, TX

**Nelson Farrier**
Teacher
Hamlin Middle School
Springfield, OR

**Michelle R. Foster**
Department Chair
Wayland Union
Middle School
Wayland, MI

**Rebecca Goodell**
Teacher
Reedy Creek Middle School
Cary, NC

**Mary Gromko**
Science Supervisor K–12
Colorado Springs District 11
Colorado Springs, CO

**Randy Mousley**
Department Chair
Dean Ray Stucky
Middle School
Wichita, KS

**David Rodriguez**
Teacher
Swift Creek Middle School
Tallahassee, FL

**Derek Shook**
Teacher
Floyd Middle Magnet School
Montgomery, AL

**Karen Stratton**
Science Coordinator
Lexington School District One
Lexington, SC

**Stephanie Wood**
Science Curriculum Specialist,
K–12
Granite School District
Salt Lake City, UT

# Online Guide

 **Get ConnectED**
connectED.mcgraw-hill.com

# ConnectED

▷ **Your Digital Science Portal**

 **Video**

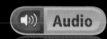 **Audio**

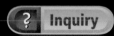 **Review**

 **Inquiry**

 **WebQuest**

See the science in real life through these exciting

Click the link and you can listen to the text while you

Try these interactive tools to help you review

Explore concepts through hands–on and virtual labs

These web-based challenges relate the concepts you're learning

The icons in your online student edition link you to interactive learning opportunities. Browse your online student book to find more.

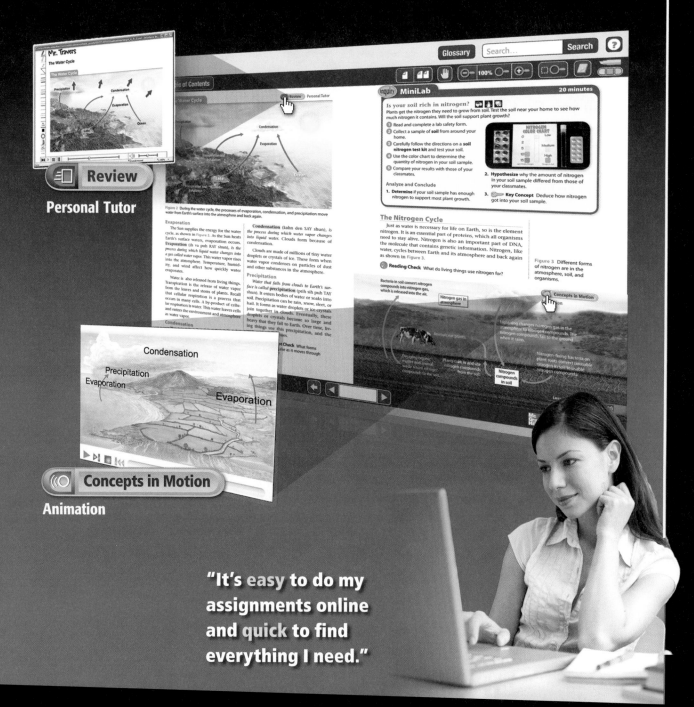

**Review**

**Personal Tutor**

**Concepts in Motion**

**Animation**

"It's easy to do my assignments online and quick to find everything I need."

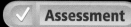

**Assessment**

Check how well you understand the concepts with online quizzes and practice questions.

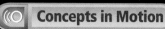

**Concepts in Motion**

The textbook comes alive with animated explanations of important concepts.

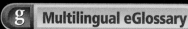

**Multilingual eGlossary**

Read key vocabulary in 13 languages.

# Treasure Hunt

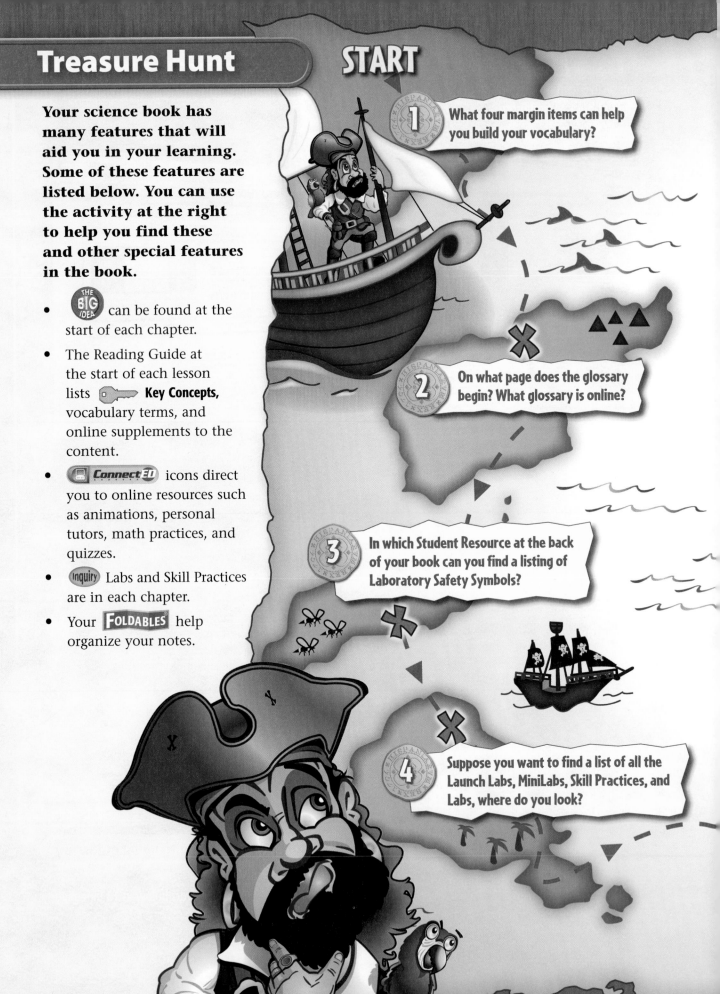

## START

Your science book has many features that will aid you in your learning. Some of these features are listed below. You can use the activity at the right to help you find these and other special features in the book.

- **BIG IDEA** can be found at the start of each chapter.

- The Reading Guide at the start of each lesson lists 🔑 **Key Concepts**, vocabulary terms, and online supplements to the content.

- **Connect ED** icons direct you to online resources such as animations, personal tutors, math practices, and quizzes.

- **Inquiry** Labs and Skill Practices are in each chapter.

- Your **FOLDABLES** help organize your notes.

**1** What four margin items can help you build your vocabulary?

**2** On what page does the glossary begin? What glossary is online?

**3** In which Student Resource at the back of your book can you find a listing of Laboratory Safety Symbols?

**4** Suppose you want to find a list of all the Launch Labs, MiniLabs, Skill Practices, and Labs, where do you look?

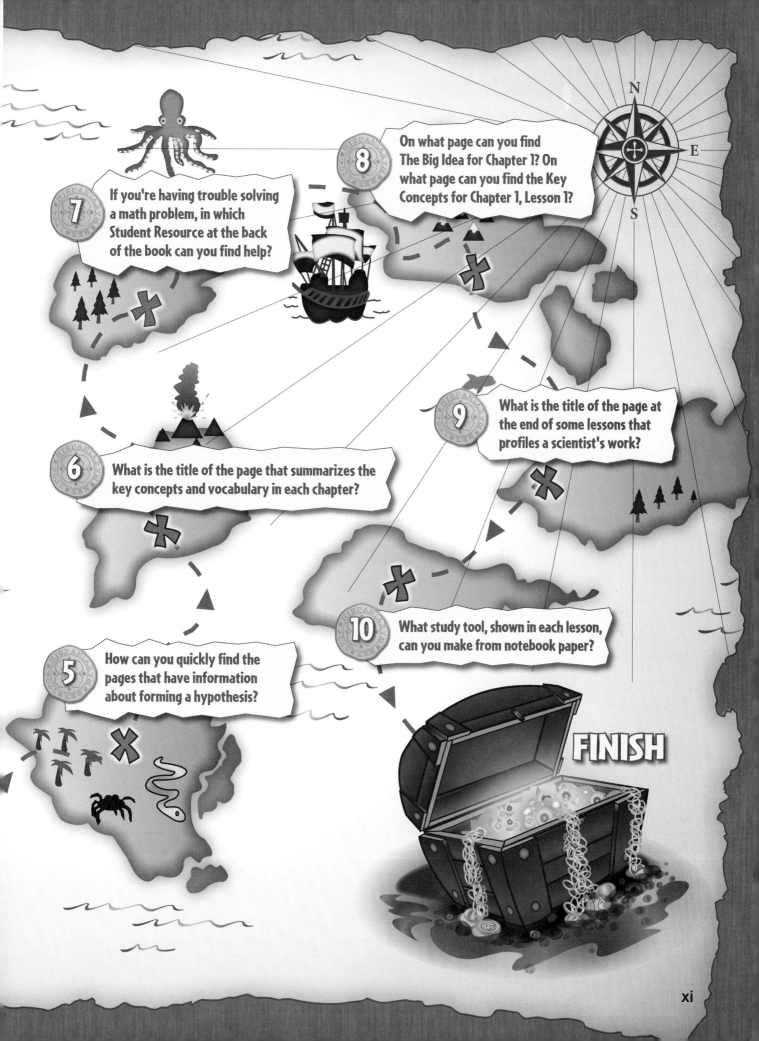

**7** If you're having trouble solving a math problem, in which Student Resource at the back of the book can you find help?

**8** On what page can you find The Big Idea for Chapter 1? On what page can you find the Key Concepts for Chapter 1, Lesson 1?

**9** What is the title of the page at the end of some lessons that profiles a scientist's work?

**6** What is the title of the page that summarizes the key concepts and vocabulary in each chapter?

**5** How can you quickly find the pages that have information about forming a hypothesis?

**10** What study tool, shown in each lesson, can you make from notebook paper?

FINISH

# Table of Contents

# Table of Contents

## Student Resources

# Inquiry

## Inquiry Launch Labs

## Inquiry MiniLabs

## Inquiry Skill Practice

## Inquiry Labs

## Features

### CAREERS in SCIENCE

# Unit 3

**100 B.C.** ⚡ **1700** **1875** **1900**

**350–341 B.C.**
Greek philosopher Aristotle classifies organisms by grouping 500 species of animals into eight classes.

**1735**
Carl Linnaeus classifies nature within a hierarchy and divides life into three kingdoms: mineral, vegetable, and animal. He uses five ranks: class, order, genus, species and variety. Linnaeus's classification is the basis of modern biological classification.

**1859**
Charles Darwin publishes *On the Origin of Species,* in which he explains his theory of natural selection.

**1866**
German biologist Ernst Haeckel coins the term *ecology.*

1950

2000

1969
American ecologist Robert Whittaker is the first to propose a five-kingdom taxonomic classification of the world's biota. The five kingdoms are Animalia, Plantae, Fungi, Protista and Monera.

1973
Konrad Lorenz, Niko Tinbergen, and Karl von Frisch are jointly awarded the Nobel Prize for their studies in animal behavior.

1990
Carl Woese introduces the three-domain system that groups cellular life-forms into Archaea, Bacteria, and Eukaryote domains.

 Inquiry
Visit ConnectED for this unit's STEM activity.

Unit 3  •  **369**

# Graphs

Polar bears are one of the largest land mammals. They hunt for food on ice packs that stretch across the Arctic Ocean. Recently, ice in the Arctic has not been as thick as it has been in the past. In addition, the ice does not cover as much area as it used to, making it difficult for polar bears to hunt. Scientists collect data about how these changes in the ice affect polar bear populations. One well-studied population of polar bears is on Wrangel Island, Russia, shown in **Figure 1.** Scientists collect and study data on polar bears to draw conclusions and make predictions about a possible polar bear extinction. Scientists often use graphs to better understand data. A **graph** is a type of chart that shows relationships between variables. Scientists use graphs to visually organize and summarize data. You can use different types of graphs to present different kinds of data.

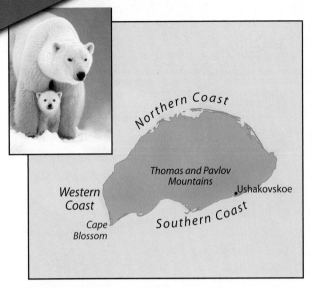

**Figure 1** Scientists collect data about polar bears on Wrangel Island, Russia.

## Types of Graphs

### Bar Graphs

The horizontal *x*-axis on a bar graph often contains categories rather than measurements. The heights of the bars show the measured quantity. For example, the *x*-axis on this bar graph contains different locations on Wrangel Island. The heights of the bars show how many bears researchers observed. The different colors show the age categories of polar bears. Where were ten adult polar bears observed?

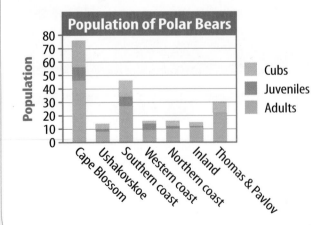

### Circle Graphs

A circle graph usually illustrates the percentage of each category of data as it relates to the whole. This circle graph shows the percentage of different age categories of polar bears on Wrangel Island. Adults, shown by the blue color, make up the largest percentage of the total population. This circle graph contains similar data to the bar graph but presents it in a different way. What percentage of the total polar bear population are cubs?

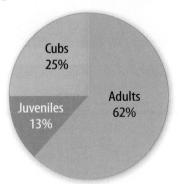

## Inquiry MiniLab

### What can graphs tell you about polar bears?

A colleague gives you some data she collected about polar bears on Wrangel Island. She observed the condition of bears near Cape Blossom and classified the bears as starving, average, or healthy. She also recorded the age category of each bear. What can you learn by graphing these data?

1 Make a bar graph of the number of bears in each category that are starving, in average condition, or healthy.

2 Add the numbers of starving bears. Add the total number of bears. Divide the number of starving bears by the total number of bears and multiply by 100 to calculate the percentage of starving bears. Repeat the calculations to find the percentages of average-condition and healthy bears. Make a circle graph showing the different conditions of the bears. For more information on how to make circle graphs, go to the Science Skill Handbook in the back of your book.

|  | Starving | Average | Healthy |
|---|---|---|---|
| Adult | 3 | 11 | 14 |
| Juvenile | 4 | 33 | 13 |
| Cub | 3 | 12 | 6 |

### Analyze and Conclude

1. **Analyze** On your bar graph, indicate how you can tell which age category of bears is the healthiest.

2. **Determine** What group of bears do you think left the most walrus carcasses? Explain.

## Line Graphs

A line graph helps you analyze how a change in one variable affects another variable. Scientists on Wrangel Island counted all the polar bears on the island each year for 10 years. They plotted each year of the survey on the horizontal *x*-axis and the bear population on the vertical *y*-axis. The population decreased between years 2 and 4. It increased between years 6 and 8. How did the population change during the last three years of the survey?

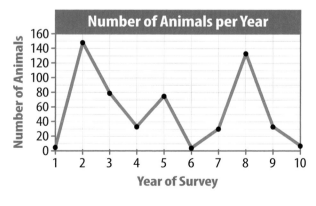

## Double Line Graphs

You can use a double line graph to compare relationships of two sets of data. The blue line represents the population of polar bears. The orange line represents the number of walrus carcasses found on Wrangel Island. You can see that the blue and orange lines follow a similar pattern. This tells scientists that these two sets of data are related. Walruses are an important food source for polar bears on the island.

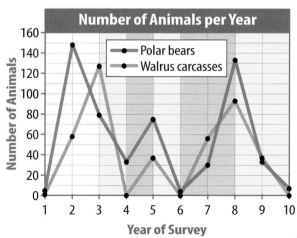

# Chapter 11

# Animal Diversity

**THE BIG IDEA**

What are the major groups of animals, and how do they differ?

**Inquiry** **Are these animals?**

What are the blue structures attached to these underwater rocks? Did someone spill paint on a clump of algae? This is a colony of animals called tunicates (TEW nuh kayts), also known as sea squirts. Believe it or not, they are classified in the same phylum as humans.

- What characteristics do you think tunicates have in common with other animals?

- How do tunicates differ from other animals?

# Get Ready to Read

## What do you think?

Before you read, decide if you agree or disagree with each of these statements. As you read this chapter, see if you change your mind about any of the statements.

**1** All animals digest food.

**2** Corals and jellyfish belong to the same phylum.

**3** Most animals have backbones.

**4** All worms belong to the same phylum.

**5** All chordates have backbones.

**6** Reptiles have three-chambered hearts.

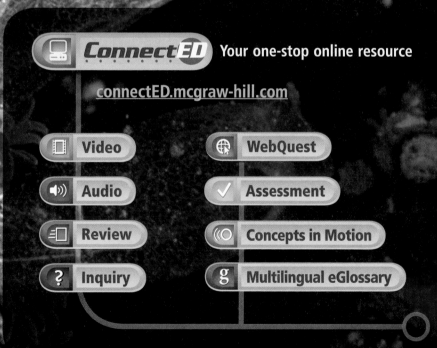

**ConnectED** Your one-stop online resource

connectED.mcgraw-hill.com

- Video
- WebQuest
- Audio
- Assessment
- Review
- Concepts in Motion
- Inquiry
- Multilingual eGlossary

# Lesson 1

## Reading Guide

**Key Concepts**
ESSENTIAL QUESTIONS

- What characteristics do all animals have?
- How are animals classified?

**Vocabulary**

**vertebrate** p. 376

**invertebrate** p. 376

**radial symmetry** p. 377

**bilateral symmetry** p. 377

**asymmetry** p. 377

g **Multilingual eGlossary**

# What defines an animal?

## Inquiry A Pair of Leaves?

What kind of organism is shown in this photo? Although they might look like leaves, they are butterflies, a type of animal. What makes leaves and butterflies different? All animals share some characteristics that leaves do not.

### What does an animal look like?

Have you ever seen an animal that looks like a vase? How about an animal that looks just like a twig? There are even some animals that look like alien spaceships. The forms that animals take are almost as varied as your imagination.

1. Look at a **photograph of an animal.** Without showing the picture to your partner, describe the animal in as much detail as possible.

2. Have your partner draw the animal using your description as a guide.

3. Compare the drawing to the photograph.

**Think About This**

1. Could someone looking at the drawing identify it as the same animal in the photograph? Why or why not?

2. 🔑 **Key Concept** What characteristics do you think you and the animal you described have in common?

## Animal Characteristics

When you look at an animal, what do you expect to see? Would you expect every animal to have legs and eyes? Ants and birds have legs, but the snake in **Figure 1** does not. Snails, spiders, and many other animals have eyes, but jellyfish do not. Although animals have many traits that make them unique, all animals have certain characteristics in common. Members of the Kingdom Animalia have the following characteristics:

- Animals are multicellular and eukaryotes.

- Animal cells are specialized for different functions, such as digestion, reproduction, vision, or taste.

- Animals have a protein, called collagen (KAHL uh juhn), that surrounds the cells and helps them keep their shape.

- Animals get energy for life processes by eating other organisms.

- Animals, such as the snake in **Figure 1,** digest their food.

In addition to the characteristics above, most animals reproduce sexually and are capable of movement at some point in their lives.

🔑 **Key Concept Check** What characteristics do all animals have?

**Figure 1** The snake began digesting its prey even before it finished swallowing.

# Animal Classification

Scientists have described and named more than 1.5 million species of animals. Every year, thousands more are added to that number. Many scientists estimate that Earth is home to millions of animal species that no one has discovered—at least, not yet. What might happen if you discovered an animal no one else had ever seen? How would you begin to classify it?

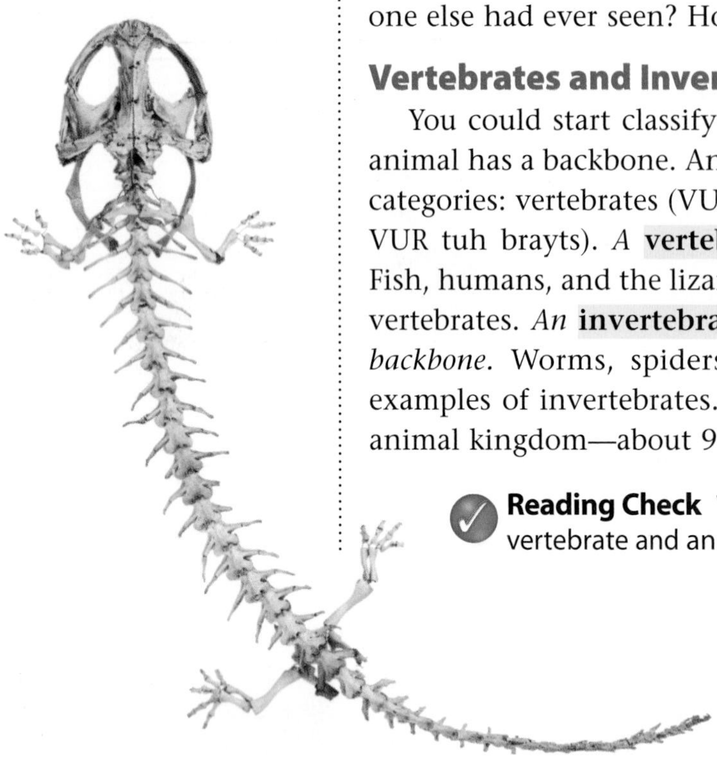

## Vertebrates and Invertebrates

You could start classifying an animal by finding out if the animal has a backbone. Animals can be grouped into two large categories: vertebrates (VUR tuh brayts) and invertebrates (ihn VUR tuh brayts). *A* **vertebrate** *is an animal with a backbone.* Fish, humans, and the lizard shown in **Figure 2** are examples of vertebrates. *An* **invertebrate** *is an animal that does not have a backbone.* Worms, spiders, snails, crayfish, and insects are examples of invertebrates. Invertebrates make up most of the animal kingdom—about 95 percent.

**Figure 2** A backbone, or spine, is part of a vertebrate's internal skeleton.

✓ **Reading Check** What is the difference between a vertebrate and an invertebrate?

---

## Inquiry MiniLab
**15 minutes**

### What is this animal?

Biologists use many characteristics to classify animals. A dichotomous key helps you identify animals based on differences in characteristics. Use the dichotomous key to identify different animals.

① Obtain a set of **animal pictures.** Follow the dichotomous key at right to identify each animal in the set.

| | |
|---|---|
| **1a.** *has a backbone* | *puffer fish* |
| **1b.** *does not have a backbone* | *go to step 2* |
| **2a.** *has bilateral symmetry* | *nudibranch* |
| **2b.** *has radial symmetry* | *go to step 3* |
| **3a.** *has spines* | *go to step 4* |
| **3b.** *does not have spines* | *ctenophore* |
| **4a.** *spherical shape* | *sea urchin* |
| **4b.** *cylindrical shape* | *sea cucumber* |

### Analyze and Conclude

1. **Observe** How many steps were there in the dichotomous key? How many organisms did you identify? What is the relationship between the number of steps in the dichotomous key and the number of animals identified?

2. **Explain** Were you uncertain about the identification of any animal? How did you decide on its identification?

3. 🔑 **Key Concept** What characteristics did you use to classify each organism?

---

**Figure 3** Animals can be classified as having radial symmetry, bilateral symmetry, or asymmetry.

✅ **Visual Check** What kind of symmetry does a bird have?

Radial symmetry

Bilateral symmetry

Asymmetry

## Symmetry

Another step you could take to classify an animal is to determine what kind of symmetry it has. As shown in **Figure 3,** symmetry describes an organism's body plan. Symmetry can help identify the phylum to which an animal belongs.

*An animal with* **radial symmetry** *can be divided into two parts that are nearly mirror images of each other anywhere through its central axis.* A radial animal has a top and a bottom but no head or tail. It can be divided along more than one plane and still have two nearly identical halves. Examples include jellyfish, sea stars, and sea anemones.

*An animal with* **bilateral symmetry** *can be divided into two parts that are nearly mirror images of each other.* Examples include birds, mammals, reptiles, worms, and insects.

*An animal with* **asymmetry** *cannot be divided into any two parts that are nearly mirror images of each other.* An asymmetrical animal, such as the sponge in **Figure 3,** does not have a symmetrical body plan.

✅ **Reading Check** What is bilateral symmetry?

**WORD ORIGIN** ··········

bilateral
from Latin *bi-*, means "two" and *latus*, means "side"

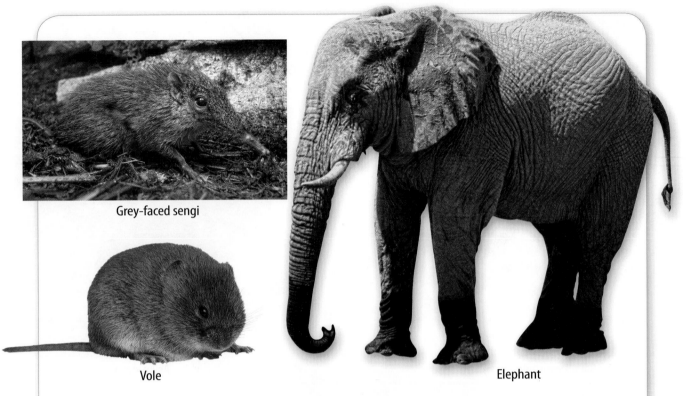

Grey-faced sengi

Vole

Elephant

Figure 4 The grey-faced sengi was first observed in Africa in 2006. Sengis look like voles, but molecular evidence shows that they are more closely related to elephants.

## Molecular Classification

Molecules such as DNA, RNA, and proteins in an animal's cells also can be used for classification. For example, scientists can compare the DNA from two animals to determine if they are related. The more similar the DNA, the more closely the animals are related.

Molecular classification has led to new discoveries about relationships among species. Scientists used to classify the grey-faced sengi shown in **Figure 4** as a close relative of shrews and voles. Recently, molecular evidence has shown that sengis are more closely related to elephants and aardvarks.

 **Key Concept Check** How are animals classified?

## Major Phyla

Scientists classify the members of the animal kingdom into as many as 35 phyla (singular, phylum). The nine major phyla, shown in **Figure 5,** contain 95–99 percent of all animal species. Animals belonging to the same phylum have similar body structures and other characteristics. For example, all sponges (the phylum Porifera [puh RIH fuh ruh]) have asymmetry, and their cells do not form tissues. Only one animal phylum, Chordata (kor DAH tuh), contains vertebrates, also shown in **Figure 5.** The other major phyla contain only invertebrates.

**FOLDABLES®**

Make a small horizontal four-door book. Leave a 1-cm space between the tabs. Draw arrows and label the book as shown. Use it to record your notes about the classification of animals.

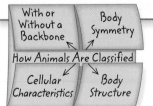

With or Without a Backbone | Body Symmetry

How Animals Are Classified

Cellular Characteristics | Body Structure

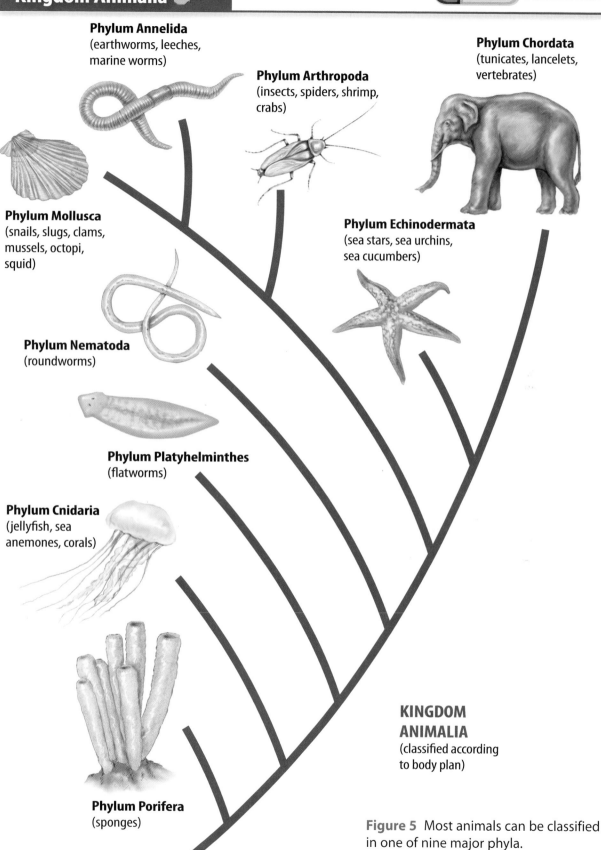

**Phylum Annelida**
(earthworms, leeches, marine worms)

**Phylum Arthropoda**
(insects, spiders, shrimp, crabs)

**Phylum Chordata**
(tunicates, lancelets, vertebrates)

**Phylum Mollusca**
(snails, slugs, clams, mussels, octopi, squid)

**Phylum Echinodermata**
(sea stars, sea urchins, sea cucumbers)

**Phylum Nematoda**
(roundworms)

**Phylum Platyhelminthes**
(flatworms)

**Phylum Cnidaria**
(jellyfish, sea anemones, corals)

**KINGDOM ANIMALIA**
(classified according to body plan)

**Phylum Porifera**
(sponges)

**Figure 5** Most animals can be classified in one of nine major phyla.

✅ **Visual Check** What are the major phyla of animals?

# Lesson 1 Review

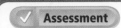 

## Visual Summary

All animals share a series of characteristics.

Animals can be classified in several ways.

Animal classifications are always changing based on advanced technology.

**FOLDABLES®**

Use your lesson Foldable to review the lesson. Save your Foldable for the project at the end of the chapter.

## What do you think NOW?

You first read the statements below at the beginning of the chapter.

**1.** All animals digest food.

**2.** Corals and jellyfish belong to the same phylum.

Did you change your mind about whether you agree or disagree with the statements? Rewrite any false statements to make them true.

## Use Vocabulary

1 **Define** *asymmetry*.

2 **Distinguish** between vertebrate and invertebrate animals.

3 **Compare and contrast** radial symmetry and bilateral symmetry.

## Understand Key Concepts

4 **List** the characteristics that all animals have in common.

5 Which characteristic applies to a horse?
   **A.** asymmetry          **C.** spherical
   **B.** invertebrate       **D.** vertebrate

## Interpret Graphics

6 **Classify** each object below as having bilateral symmetry, radial symmetry, or asymmetry.

7 **Summarize Information** Copy the graphic organizer below, and use it to summarize the ways animals can be separated into groups.

## Critical Thinking

8 **Develop** a series of instructions that could be used to determine if an animal should be classified in the phylum Arthropoda, Echinodermata, or Chordata.

9 **Analyze** how the classification of the grey-faced sengi changed over time. How might technological advances change how other animals are classified?

# A Family Tree for Bats

*Meet Nancy Simmons, a taxonomist who identifies bats.*

When most people are going to bed, taxonomist Nancy Simmons is going to work. She's off to capture bats in a dense rain forest of South America. Because bats are most active at night, she and her team from the American Museum of Natural History work from dusk until dawn. They must capture, identify, and release the bats while it's dark.

Taxonomists study animals to see how they are related to each other and use that information to classify them. To classify a bat, Simmons carefully examines its body. She looks at characteristics such as wing size, fur color, and the shape of the bat's teeth. These characteristics help her classify each bat into a family or a group that shares physical features and behaviors.

In 1999 Dr. Simmons added a new member to the bat family tree. In the rain forest in Peru, her team discovered a species they named *Micronycteris matses,* the Matses' big-eared bat. Like other species in its genus, *M. matses* is small with large round ears, a long snout, and a fold of skin on its nose called a nose-leaf. *M. matses* is unique, however, because of its combination of dark brown fur, medium body size, small bottom front teeth, and short fur around its ears.

Dr. Simmons is looking for links between *M. matses* and other bat species. She compares their bodies, behavior, and even their DNA. Her goal is to create a family tree for all bats. With over a thousand species of bats worldwide, Dr. Simmons has plenty of work still to do.

## All Kinds of Bats

Bats live on every continent except Antarctica, in areas ranging from tropical rain forests to chilly mountaintops. They also have an amazing variety of shapes and sizes. With over 1,100 species, bats make up one-fifth of the world's mammals.

**Simmons holds a bat that she caught in her net.**

## It's Your Turn

**RESEARCH** Investigate a species of bat in your area. Record where it lives in the environment and its characteristics, such as wingspan, fur color, and weight. With a partner, compare your bats. What do they have in common? What is different?

▲ **This species is the largest in the New World—it weighs about 150 g.**

### Reading Guide

**Key Concepts**
ESSENTIAL QUESTIONS

- What are the characteristics of invertebrates?
- How do the invertebrate phyla differ?

**Vocabulary**

**exoskeleton** p. 387

**appendage** p. 387

**g** Multilingual eGlossary

Video BrainPOP®

# Invertebrate Phyla

Inquiry **How did it get there?**

This octopus is alive and got inside the bottle by slowly pushing its soft, flexible body inside. Like many invertebrates, an octopus does not have a skeleton made of bone or other hard structures.

## What does an invertebrate look like?

Some invertebrates have features that are similar to yours, such as eyes and legs. Others have little in common with you. What do you see when you look at invertebrates close-up?

1. Read and complete a lab safety form.
2. Examine a **collection of invertebrates,** and record your observations in your Science Journal.
3. Use a **magnifying lens** to further examine the invertebrates. Record any additional observations.
4. Make a Venn diagram in your Science Journal to compare similarities and contrast differences among the invertebrates.

### Think About This

1. Which two invertebrates were the most dissimilar? Why?

2. Did you see any details using a magnifying lens that you missed by looking just with your eyes?

3. 🔑 **Key Concept** What characteristics do you think all the invertebrates you looked at have in common?

# Characteristics of Invertebrates

Can you imagine living without a backbone? Most animals do just that. As you have read, invertebrates are animals that lack a backbone. In most cases, invertebrates have no internal structures to help support their bodies. They also tend to be smaller and move more slowly than vertebrates. As shown in **Figure 6,** over 95 percent of all animal species that have been recorded are invertebrates.

You probably could recognize a jellyfish or a clam if you saw one. What about an anemone or a sea cucumber? Invertebrates are a diverse group. Their physical characteristics range from the simple structures of sponges and jellyfish to the more complex bodies of worms, snails, and insects. Each invertebrate phylum contains animals with similar body plans and physical characteristics.

🔑 **Key Concept Check** What are the characteristics of invertebrates?

**ACADEMIC VOCABULARY**

internal
*(adjective)* existing inside something

**Invertebrate and Vertebrate Species**

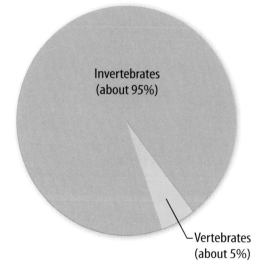

Invertebrates
(about 95%)

Vertebrates
(about 5%)

**Figure 6** 🔑 Invertebrates make up more than 95 percent of all living species on Earth.

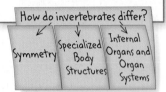

FOLDABLES®

Make a horizontal three-tab book. Draw arrows and label the tabs as shown. Use the Foldable to identify differences found within the invertebrate phyla.

How do invertebrates differ?

Symmetry | Specialized Body Structures | Internal Organs and Organ Systems

## Sponges and Cnidarians

The simplest of the invertebrates are the sponges, which belong to the phylum Porifera. All sponges share several characteristics.

All sponges are asymmetrical and have no tissues, organs, or organ systems. Their cells are specialized for capturing food, digestion, and reproduction. Other cells provide support inside the layers of the sponge. All sponges live in water, and most species live in ocean environments.

The phylum Cnidaria (ni DAR ee uh) includes jellyfish, sea anemones, hydras, and corals. Cnidarians, such as the sea anemone shown in **Figure 7,** differ from all other animals based on their unique characteristics.

Cnidarians have no organs or organ systems, but, unlike sponges, they have radial symmetry. They have a single body opening surrounded by tentacles. Simple tissues, including muscles, nerves, and digestive tissue, enable cnidarians to survive by moving, reacting to stimuli, and digesting food. They have specialized cells, called nematocysts (NE mah toh sihsts), that are used for defense and capturing food. Similar to sponges, most species of cnidarians live in ocean environments, and all live in water.

 **Reading Check** What characteristics do poriferans and cnidarians share?

**Figure 7** The tentacles of all cnidarians contain stinging structures for capturing food and defending against predators.

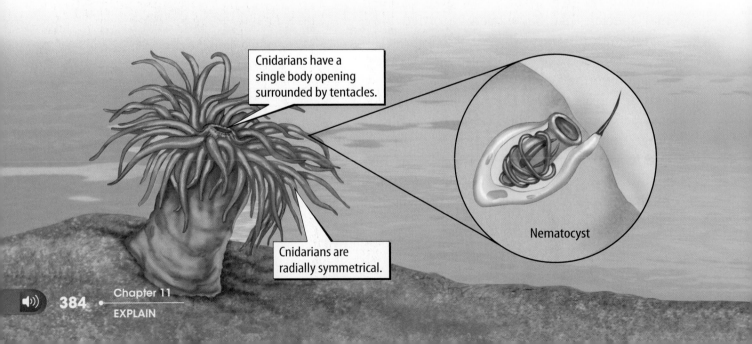

Cnidarians have a single body opening surrounded by tentacles.

Cnidarians are radially symmetrical.

Nematocyst

# Flatworms and Roundworms

Flatworms are invertebrates that belong to the phylum Platyhelminthes (pla tih hel MIHN theez). All flatworms, including the tapeworm shown in **Figure 8,** have bilateral symmetry with nerve, muscle, and digestive tissues and a simple brain. They have soft and flattened bodies that are usually only a few cells thick. The digestive system of a flatworm has only one opening: a mouth.

Flatworms live in moist environments. Most, like tapeworms, are parasites that live in or on the bodies of other organisms and rely on them for food. Others are free-living, and many live in oceans or other marine environments.

 **Reading Check** What characteristics do all flatworms share?

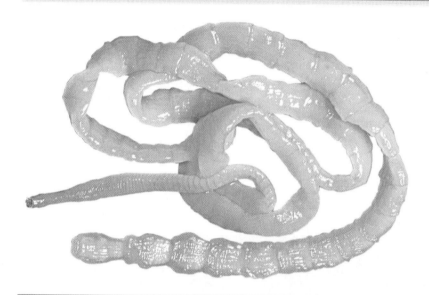

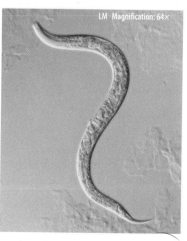

▲ **Figure 8** Most flatworm species, including this tapeworm, are parasites. They depend on other organisms for food and a place to live.

**Visual Check** How would you describe a flatworm's body?

Roundworms, also called nematodes, belong to the phylum Nematoda (ne muh TOH duh). Roundworms, like flatworms, have bilateral symmetry with nerve, muscle, and digestive tissues and a simple brain. However, unlike flatworms, their bodies are round and covered with a stiff outer covering called a cuticle. A roundworm's digestive system has two openings: a mouth and an anus. Food enters the mouth and is digested as it travels to the anus where wastes are excreted.

Roundworms live in moist environments. Some species are parasites that live in animals' digestive systems. Free-living roundworms such as the one pictured in **Figure 9** eat material such as fecal matter and dead organisms.

LM Magnification: 64×

▲ **Figure 9** Roundworms are narrow and tapered at both ends. Most species are less than 1 mm long.

**Reading Check** How do flatworms and roundworms differ?

## Mollusks and Annelids

The phylum Mollusca (mah LUS kuh) includes snails, slugs, clams, mussels, octopi, and squid. All mollusks, including the snail shown in **Figure 10,** have bilateral symmetry. Their bodies are soft, and some species have hard shells that protect their bodies. You might have seen a slug slithering along the ground after a rainstorm. Slugs are one type of mollusk without a shell.

Mollusks have digestive systems with two openings. A body cavity contains the heart, the stomach, and other organs. The mollusk circulatory system contains blood, but no blood vessels. Their nervous systems include eyes and other sensory organs as well as simple brains. Members of this phylum must remain wet and live in water or moist environments.

The phylum Annelida includes earthworms, leeches, and marine worms. Annelid worms, including the one shown in **Figure 11,** have bilateral symmetry and soft bodies. Their bodies consist of repeating segments covered with a thin cuticle. Their digestive systems have two openings. Annelids have circulatory systems that are made up of blood vessels that carry blood throughout the body. Their nervous systems include a simple brain. Annelids live in water or moist environments such as soil.

**Reading Check** What do mollusks and segmented worms have in common?

▲ **Figure 10** Snails have shells that protect their bodies.

**Figure 11** One characteristic that distinguishes annelids from other worms is their segments. ▼

**✔ Visual Check** How does this annelid differ from an earthworm?

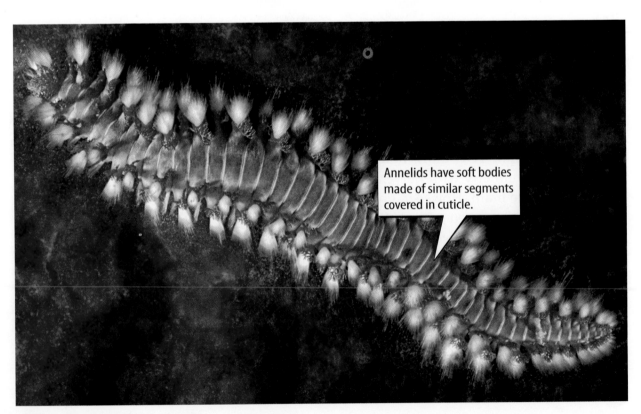

Annelids have soft bodies made of similar segments covered in cuticle.

# Arthropods

The phylum Arthropoda includes insects, spiders, shrimp, crabs, and their relatives. More species belong to this phylum than all the other animal phyla combined. There are more than 1 million identified species of arthropods.

All arthropods have bilateral symmetry. They also have **exoskeletons**—*thick, hard outer coverings that protect and support animals' bodies.* Arthropods have several pairs of jointed appendages. *An* **appendage** *is a structure, such as a leg or an arm, that extends from the central part of the body.* The body parts of arthropods are segmented and specialized for different functions such as flying and eating. Unlike many of the other animals you have read about so far, arthropods live in almost every environment on Earth.

 **Reading Check** What do exoskeletons do?

## Insects

The largest order of arthropods is the insects, which includes the stag beetle shown in **Figure 12.** All insect species have three pairs of jointed legs, three body segments, a pair of antennae, and a pair of compound eyes. Many species also have one or two pairs of wings.

There are 16 major groups of insects. However, most insect species belong to one of five groups. Beetles form the largest group of insects. About 40 percent of all known species of insects are beetles.

**WORD ORIGIN**

**appendage**
from Latin *appendere,* means "to cause to hang from"

**Figure 12** A stag beetle has characteristics common to all insect species.

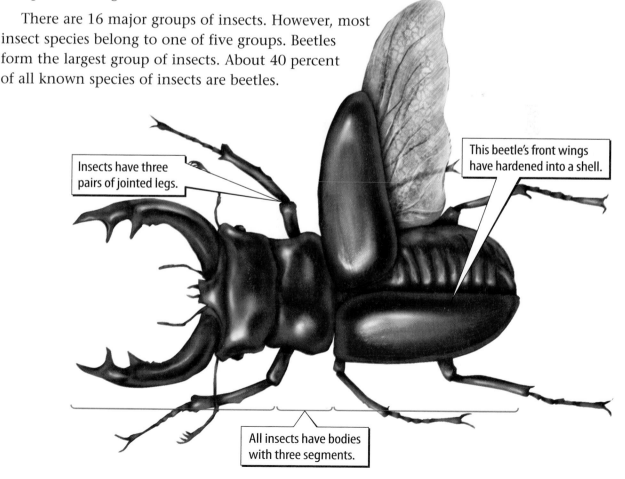

Insects have three pairs of jointed legs.

This beetle's front wings have hardened into a shell.

All insects have bodies with three segments.

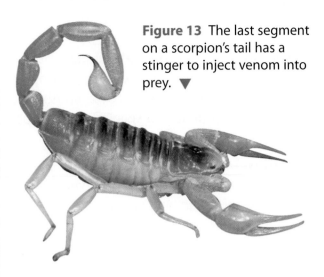

**Figure 13** The last segment on a scorpion's tail has a stinger to inject venom into prey. ▼

## Inquiry MiniLab  20 minutes

### How does your arm move?

Both arthropods and mammals have jointed appendages. Try doing some simple tasks without bending your appendages to understand how useful jointed appendages are.

1 Using **newspaper** and **masking tape,** wrap your partner's arm at the elbow so he or she cannot bend it.

2 Ask your partner to perform the tasks in the data table below. Record your observations and your partner's experiences.

| Task | Completed? (yes/no) | How was behavior changed? |
|---|---|---|
| Walk 5 m. | | |
| Take a drink of water. | | |
| Lay down on the ground and then stand up. | | |

### Analyze and Conclude

1. **Summarize** Rank the tasks in order from hardest to easiest to perform without jointed appendages. What made the tasks harder to perform?

2. **Infer** What activities that you must perform in order to survive are impossible without jointed appendages?

3. 🔑 **Key Concept** Explain how jointed appendages are necessary for arthropods to survive.

## Arachnids

Spiders, ticks, and scorpions, such as the one shown in **Figure 13,** are arachnids (uh RAK nudz). All arachnids have four pairs of jointed legs and two body segments. They do not have antennae or wings.

## Crustaceans

Crabs, shrimp, lobsters, and their close relatives are crustaceans (krus TAY shunz). All crustaceans have one or two pairs of antennae. They also have jointed appendages in the mouth area that are specialized for biting and crushing food. Many people like to eat crustaceans, including lobsters and crabs, such as the one shown in **Figure 14.**

✓ **Reading Check** How do arachnids and crustaceans differ?

▲ **Figure 14** Like most crustaceans, this crab's eyes are on stalks.

# Echinoderms

The phylum Echinodermata (ih kin uh DUR muh tuh) includes sea stars, sea cucumbers, and sea urchins, such as the one shown in **Figure 15**. *Echinoderm* (ih KI nuh durm) means "spiny skin." Echinoderms have some unique features that are not in any of the other invertebrate phyla. They also are more closely related to vertebrates than to any other phyla.

All echinoderms have radial symmetry. Unlike any other phyla, echinoderms have hard plates embedded in the skin that support the body. Thousands of small, muscular, fluid-filled tubes, called tube feet, enable them to move and feed. They also have complete digestive systems including a mouth and an anus. Echinoderms live only in oceans. However, some can survive out of the water for short periods during low tides.

 **Key Concept Check** How do the invertebrate phyla differ?

**Figure 15** Sea urchins are one type of echinoderm.

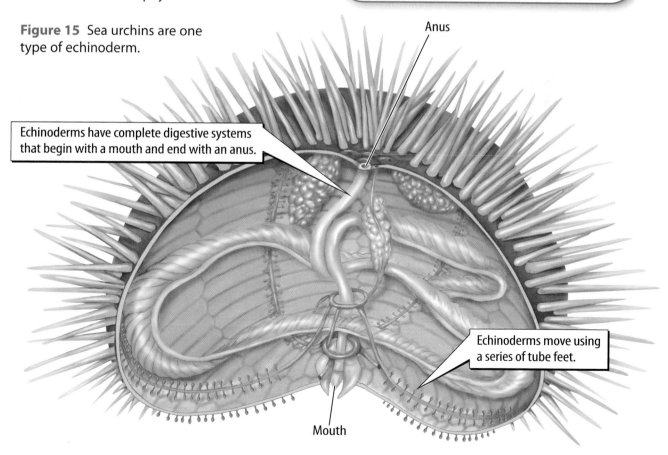

Anus

Echinoderms have complete digestive systems that begin with a mouth and end with an anus.

Echinoderms move using a series of tube feet.

Mouth

# Lesson 2 Review

## Visual Summary

Most invertebrates have no internal structures that support their bodies.

There are more arthropods than members of any other phyla.

The largest order of arthropods is the insects.

**FOLDABLES**

Use your lesson Foldable to review the lesson. Save your Foldable for the project at the end of the chapter.

## What do you think NOW?

You first read the statements below at the beginning of the chapter.

**3.** Most animals have backbones.

**4.** All worms belong to the same phylum.

Did you change your mind about whether you agree or disagree with the statements? Rewrite any false statements to make them true.

## Use Vocabulary

**1** **Define** *exoskeleton*.

**2** **Distinguish** between the phylum Platyhelminthes and the phylum Nematoda.

**3** **Use the term** *appendage* in a sentence.

## Understand Key Concepts

**4** Which phylum contains asymmetrical invertebrates that have no tissues?
  A. Annelida      C. Echinodermata
  B. Cnidaria      D. Porifera

**5** **Describe** the characteristics of the phylum that contains invertebrates with wings.

## Interpret Graphics

**6** **Summarize Information**  Copy the table below, and fill in the features common to the members of each invertebrate phylum.

| Phylum | Characteristics | Example |
|---|---|---|
| Porifera | | |
| Cnidaria | | |
| Platyhelminthes | | |
| Annelida | | |
| Nematoda | | |
| Arthropoda | | |
| Echinodermata | | |

## Critical Thinking

**7** **Hypothesize** how a digestive system with two openings would enable an organism to absorb more nutrients than a digestive system with one opening.

## Math Skills

Review
———— Math Practice ————

**8** About 11,000 species of Lepidoptera (butterflies and moths) have been identified in the United States. Only 679 of them are butterflies. What percentage of the Lepidoptera species in the United States are butterflies?

# How do you build a dichotomous key?

### Materials

invertebrates

magnifying
lens

### Safety

A dichotomous key helps you classify animals based on their characteristics. *Dichotomous* means "divided in two parts." Each step of the key has two choices. You choose the one that applies to the animal you are studying, and it directs you to the next set of choices. By picking the best choices for an animal's characteristics, you can classify animals in a list of possibilities.

## Learn It

Sorting objects into groups based on common features is called **classifying.** When classifying, first observe the objects being classified. Then select one feature that is shared by some, but not all, of the objects. Place all the members that share a feature into a subgroup. You can classify members into smaller and smaller subgroups based on characteristics.

## Try It

① Read and complete a lab safety form.

② Study your invertebrate collection. You may want to use a magnifying lens. Step 1 of your dichotomous key is to divide your collection into two groups based on a characteristic. Make a table like the one shown below.

| Step | Characteristic | "Go to"/ Identity |
|------|----------------|-------------------|
| 1. | legs present | step 2 |
|    | legs absent | step 3 |

③ Now think about the subgroup of animals that have the characteristic in step 1. Divide these animals into two smaller subgroups based on another characteristic. Enter this choice in step 2 of your dichotomous key.

④ Suppose only one animal in your collection falls into a subgroup. Place the identity of the animal in the right column of the table.

| Step | Characteristic | "Go to"/ Identity |
|------|----------------|-------------------|
| 1. | legs present | step 2 |
|    | legs absent | step 3 |
| 2. | wings present | step 4 |
|    | wings absent | pavement ant (*Tetramorium caespitum*) |

⑤ Repeat steps 3–5 until the animals are each in their own subgroup and the dichotomous key leads you to the identity of each animal.

## Apply It

⑥ **Identify** Remove the labels from your animal collection. Trade your collection and your dichotomous key with a classmate. Identify all the animals in your classmate's collection using his or her key. Check your answers.

⑦ 🔑 **Key Concept** What characteristics did all the animals you identified have in common?

# Lesson 3

## Phylum Chordata

## Reading Guide

### Key Concepts 🔑
**ESSENTIAL QUESTIONS**

- What are the characteristics of all chordates?

- What are the characteristics of all vertebrates?

- How do the classes of vertebrates differ?

### Vocabulary
**notochord** p. 393

**chordate** p. 393

 **Multilingual eGlossary**

 **Video** **BrainPOP®**

### Inquiry One of a Kind?

Several different types of animals come to this watering hole to get a drink. These elephants, antelopes, and birds look very different, but would you guess that they were all related? All of these animals belong to the phylum Chordata and share several characteristics.

**10 minutes**

## How can you model a backbone?

All vertebrates have backbones. Most backbones are made out of a stack of short bones called vertebrae. Some vertebrae are shaped like discs with holes in the center. The largest structure passing through the center of the stack of vertebrae is the spinal cord. Between each of the vertebrae are padlike structures, called discs, that cushion the bones. Try building a model of a backbone.

1 Read and complete a lab safety form.

2 Obtain **pasta wheels, circular gummy candies,** and a **chenille stem.** ⚠ Do not eat the lab materials.

3 Assemble the materials to make a model of a backbone.

4 Gently bend and move your model backbone. Observe how the parts move and interact with each other.

### Think About This

1. When you bend your model backbone, how are the vertebrae, the discs, and the spinal cord affected?

2. When you compress your model backbone, how are the vertebrae, the discs, and the spinal cord affected?

3. 🔑 **Key Concept** How do you think the structure of the backbone provides advantages to the body plan of vertebrates?

## Characteristics of Chordates

Recall that one way to classify an animal is to check for a backbone and that animals with backbones are called vertebrates. Another way to classify animals is to look for the four characteristics of a chordate (KOR dayt). *A* **chordate** *is an animal that has a notochord, a nerve cord, a tail, and structures called pharyngeal* (fer IN jee ul) *pouches at some point in its life.* In vertebrates, these characteristics are present only during embryonic development. *A* **notochord** *is a flexible, rod-shaped structure that supports the body of a developing chordate.* The nerve cord develops into the central nervous system. The pharyngeal pouches are between the mouth and the digestive system.

Most chordates are vertebrates, but the chordates also include two groups of invertebrates: tunicates and lancelets (LAN sluhts), shown in **Figure 16.** Invertebrate chordates are rarely more than a few centimeters long and live in salt water. In vertebrate chordates, such as humans, the notochord develops into a backbone during the growth of an embryo.

🔑 **Key Concept Check** What are the characteristics of chordates?

**Figure 16** Lancelets can swim but spend most of their lives almost completely buried in sand.

A lancelet is one type of invertebrate that has a notochord.

**FOLDABLES**®

Make a vertical five-tab book. Label the tabs as shown. Use the Foldable to identify specific characteristics and examples of vertebrates.

Fish

Amphibians

Reptiles

Birds

Mammals

## Characteristics of Vertebrates

Recall that all vertebrates have a backbone, also called a spinal column or spine. The backbone is a series of structures that surround and protect the nerve cord, or spinal cord. The spinal cord connects all the nerves in the body to the brain. Bones that form a backbone are called vertebrae (VUR tuh bray). If you gently touch the back of your neck, the bones you feel are some of your vertebrae.

Vertebrates have well-developed organ systems. All vertebrates have digestive systems with two openings, circulatory systems that move blood through the body, and nervous systems that include brains. The five major groups of vertebrates are fish, amphibians, reptiles, birds, and mammals.

 **Key Concept Check** What are the characteristics of all vertebrates?

## Fish

Most fish spend their entire lives in water. They have two important characteristics in common: gills for absorbing oxygen gas from water and paired fins for swimming. Fish are grouped into one of three classes.

Hagfish and lampreys lack jaws and are in a group called jawless fish. Sharks, such as the one shown in **Figure 17,** skates, and rays are cartilaginous fish. They have skeletons made of a tough, fibrous tissue called cartilage (KAR tuh lihj). Both jawless and cartilaginous fish have internal structures made of cartilage.

Trout, guppies, perch, tuna, mackerel, and thousands of other species do not have cartilaginous skeletons. Instead, they have bones and are grouped together as bony fish.

**Figure 17** Like all fish, sharks have gills and fins.

Structures called gills enable fish to absorb oxygen from the water.

## Amphibians

Frogs, toads, and salamanders belong to the class Amphibia, as shown in **Figure 18.** Most **amphibians** spend part of their lives in water and part on land. Their bodies change as they grow older. In many species, the young have different body forms than the adults do.

Amphibians have skeletons made of bone and have legs for movement. Their skin is smooth and moist, and their hearts have three chambers. Amphibians lay eggs that do not have hard protective coverings, or shells. Their eggs must be laid in moist environments, such as ponds. Young live in water and have gills; most adults develop lungs and live on land.

 **Reading Check** How do amphibians differ from fish?

**WORD ORIGIN**
**amphibian**
from Greek *amphi-*, means "of both kinds" and *bios*, means "life"

Adult amphibians have lungs and live on land.

Amphibian eggs do not have shells.

Young amphibians have gills.

**Figure 18** The body forms of many amphibians change as they grow.

**Visual Check** How does the body form of this salamander change as it grows?

**Figure 19** This lizard has a three-chambered heart and lays fluid-filled eggs.

# Reptiles

Lizards, snakes, turtles, crocodiles, and alligators belong to the class Reptilia. A leopard gecko, one example of this class, is shown in **Figure 19.**

All reptiles share several characteristics. Their skin is waterproof and covered in **scales.** Like amphibians, most reptiles have three-chambered hearts. Unlike amphibians, lizards and other reptiles have lungs throughout their lives.

Most reptiles lay fluid-filled eggs with leathery shells. Unlike amphibian eggs, reptile eggs are laid on land rather than in water. Young reptiles do not change form as they mature into adult reptiles.

✓ **Reading Check** How do reptiles differ from amphibians?

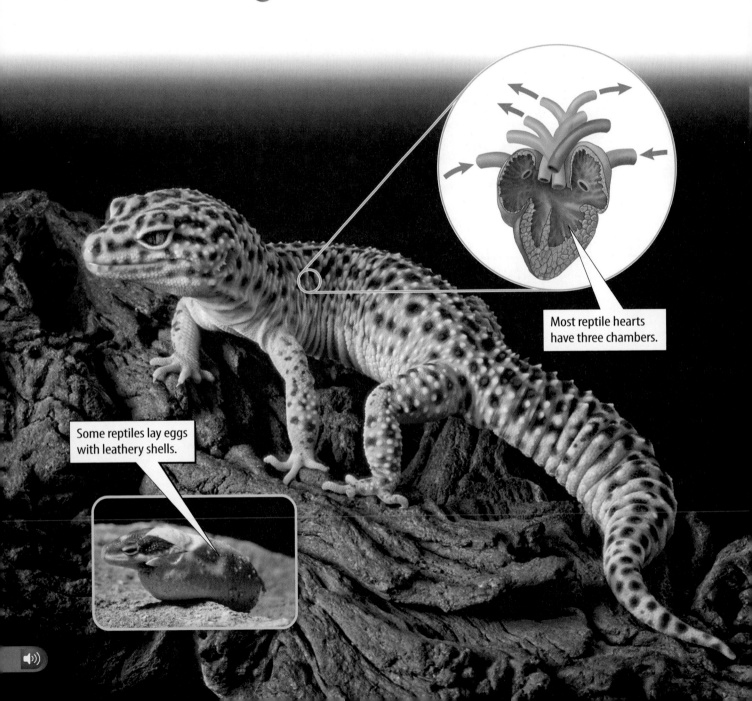

Most reptile hearts have three chambers.

Some reptiles lay eggs with leathery shells.

# Birds

All birds, including the owl shown in **Figure 20,** are in the class Aves. Many birds make nests to hold their **eggs,** and many have unique calls or songs.

Birds have lightweight bones. Their skin is covered with feathers and scales. Birds also have two legs and two wings. Many birds can fly, and they have stiff feathers that enable them to move through the air. Birds that spend a lot of time in the water have oil glands that help water roll off their feathers.

Birds have beaks and do not chew their food. Instead, their digestive systems include gizzards, organs that help grind food into smaller pieces. Their circulatory systems include four-chambered hearts. Birds also lay fluid-filled eggs with hard shells and feed and care for their young.

**Reading Check** How do birds differ from reptiles?

**REVIEW VOCABULARY** · · · · · · · · · · · · · · ·

egg
a female sex cell that forms in an ovary

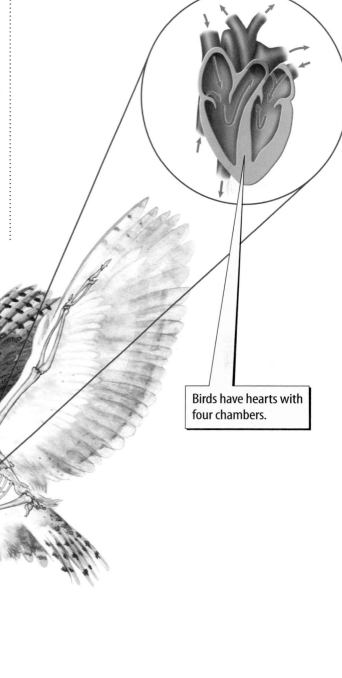

Birds have hearts with four chambers.

**Figure 20** All birds have several characteristics in common, including lightweight bones and four-chambered hearts.

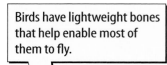

Birds have lightweight bones that help enable most of them to fly.

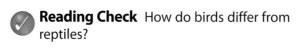

## Mammals

Dogs, cats, goats, rats, seals, whales, and humans are among the many vertebrates belonging to the class Mammalia. All mammals have hair or fur covering their bodies. As shown in **Figure 21,** they tear and chew their food using teeth. Mammals have complete digestive systems, which include a mouth and an anus, and a complex nervous system including a brain.

The most notable characteristic of mammals, however, is the presence of mammary glands. These glands produce milk that feeds young mammals. Although many mammals have live young, a few species, including the duck-billed platypus, lay eggs.

**Figure 21** Mammals have hair or fur and mammary glands.

 **Key Concept Check** How do the classes of vertebrates differ?

---

**Inquiry MiniLab**                    **15 minutes**

### Whose bones are these?

The skeletons of vertebrates are made up of bones. Bones have different characteristics depending on the animal in which they are found and their function in the body. Observe bones from different animals.

1. Read and complete a lab safety form.
2. Obtain a collection of **bones.**
3. Examine the shape, the texture, the mass, and the size of the bones.
4. Copy the table below in your Science Journal. Record your observations in the table. Add extra columns if you notice other characteristics you would like to record.

**Analyze and Conclude**

1. **Compare** What traits did all the bones you observed share?

2. **Contrast** What was the biggest difference among the bones you observed?

3. **Key Concept** Use your observations to identify bones from two different classes of vertebrates.

| Bone | Shape | Texture | Mass | Size | Other Observations |
|------|-------|---------|------|------|--------------------|
| 1 | | | | | |
| 2 | | | | | |
| 3 | | | | | |
| 4 | | | | | |

# Lesson 3 Review

## Visual Summary

Most chordates are vertebrates.

Vertebrates have well-developed organ systems including digestive systems with two openings, circulatory systems that move blood through the body, and nervous systems including brains.

Mammals produce milk to feed their young.

**FOLDABLES®**

Use your lesson Foldable to review the lesson. Save your Foldable for the project at the end of the chapter.

## What do you think NOW?

You first read the statements below at the beginning of the chapter.

**5.** All chordates have backbones.

**6.** Reptiles have three-chambered hearts.

Did you change your mind about whether you agree or disagree with the statements? Rewrite any false statements to make them true.

## Use Vocabulary

**1** **Distinguish** between reptiles and amphibians.

**2** **Define** *notochord*.

## Understand Key Concepts

**3** Which characteristic is common to all chordates?
- **A.** bones
- **B.** fur
- **C.** lungs
- **D.** notochord

**4** **List** the characteristics common to all fish.

**5** **Compare and contrast** birds and mammals.

## Interpret Graphics

**6** **Summarize Information** Copy the table below, and fill in the features of each type of chordate.

| Type of Animal | Characteristics | Example Animals |
|---|---|---|
| Invertebrate chordates | | |
| Vertebrate chordates | | |

**7** **Analyze** To which class of vertebrates does the animal below belong? How do you know?

## Critical Thinking

**8** **Assess** Why does a backbone make vertebrates better adapted for life on land than invertebrates? Explain your answer.

**9** **Infer** What is the advantage of having bones that protect the central nerve cord?

# Design Your Own Phylum

## Materials

markers

colored pencils

In this chapter, you have learned that different types of animals have different characteristics. Some animals have radial symmetry. Others have bilateral symmetry. Still others have no symmetry at all. Vertebrates have backbones. Invertebrates do not. Your task is to design a new phylum of alien animals that has never before been described. What are the characteristics of the animals in your phylum? Remember that animals must perform tasks in order to survive, such as capturing food and reproducing. What characteristics enable the animals in your phylum to survive?

## Question

What characteristics enable you to identify animals in your new phylum? What different characteristics do animals in your phylum have?

## Procedure

1 Read and complete a lab safety form.

2 Spend time thinking about your phylum. In your Science Journal, write down characteristics that are common to all the animals in your phylum. Try writing a list of structures animals must have and functions that animals must perform as a guide for creating your phylum's characteristics. Create a name for your phylum.

3 Create five different species of animals in your phylum. Write a list of characteristics for each of these animals that make them different from each other. Name each animal.

4 On five separate pieces of paper, draw each of your animals to the best of your ability.

5 Build a dichotomous key that someone could use to identify the animals in your phylum.

6 Trade pictures of animals and dichotomous keys with a classmate. Using your classmate's dichotomous key, identify each of the animals in the drawings.

7 **Analyze** Show your classmate your identifications. Were your answers correct? If not, make modifications to your identifications. Record your changes in your Science Journal.

## Analyze and Conclude

8 **Compare** your phylum with your classmate's phylum. What characteristics did the two phyla share? Name some characteristics that were different.

9 **Evaluate** What types of characteristics were the most useful for identifying animals? What types were the hardest to use?

10 **The Big Idea** If your new phylum was included in Kingdom Animalia, where would it go? Why would it be placed in that location?

## Communicate Your Results

Suppose that you are a zoologist, and you have discovered animals in your new phylum. Prepare a press release describing your phylum. Use your pictures to illustrate the characteristics of the animals you discovered. Explain where you found your animals, how they survive in the wild, and any other information that makes your phylum interesting.

**Inquiry** Extension

Try building physical models of the animals in your phylum. Use wood, wire, clay, paint, and other sculpting materials.

## Lab Tips

☑ Try thinking about the environment where you would find your phylum to help you think of characteristics that enable it to live there.

☑ Think about the animals that might have been ancestors to phyla that exist today. What predators might the animals in your phylum have?

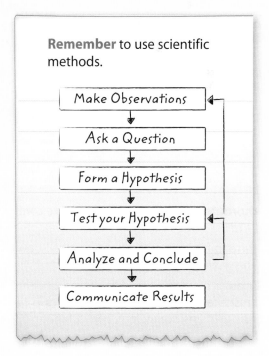

**Remember** to use scientific methods.

Make Observations
↓
Ask a Question
↓
Form a Hypothesis
↓
Test your Hypothesis
↓
Analyze and Conclude
↓
Communicate Results

# Chapter 11 Study Guide

 WebQuest

 **THE BIG IDEA**

**The major groups of animals include sponges, cnidarians, flatworms, roundworms, mollusks, segmented worms, arthropods, and chordates. They differ based on body structures and types of reproduction.**

| Key Concepts Summary 🔑 | Vocabulary |
|---|---|
| **Lesson 1: What defines an animal?** <br><br> • Animals are eukaryotic, multicellular organisms that eat other organisms, digest food, and have collagen to support cells. Most animals reproduce sexually and can move. <br><br> • Animals can be classified based on the presence of a backbone; body symmetry; the characteristics of proteins, DNA, and other molecules that make up their cells; and the kinds of body structures they possess. <br><br>  | **vertebrate** p. 376 <br> **invertebrate** p. 376 <br> **radial symmetry** p. 377 <br> **bilateral symmetry** p. 377 <br> **asymmetry** p. 377 |
| **Lesson 2: Invertebrate Phyla** <br><br> • Invertebrates have no backbone or internal skeleton, and they tend to be smaller and slower-moving than vertebrates. <br><br> • Invertebrates differ based on symmetry, presence or absence of certain types of specialized body structures, and presence or absence of specific internal organs and organ systems. <br><br> 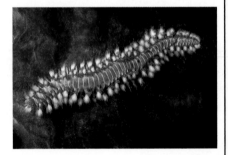 | **exoskeleton** p. 387 <br> **appendage** p. 387 |
| **Lesson 3: Phylum Chordata** <br><br> • All **chordates** have a **notochord**, a nerve cord, pharyngeal pouches, and a tail at some time during their development. <br><br> • All vertebrates have a backbone and well-developed organs and organ systems. <br><br> • The classes of vertebrates differ based on presence or absence of characteristics such as gills, fins, scales, legs, wings, fur, and eggs. <br><br>  | **notochord** p. 393 <br> **chordate** p. 393 |

## FOLDABLES® Chapter Project

Assemble your lesson Foldables as shown to make a Chapter Project. Use the project to review what you have learned in this chapter.

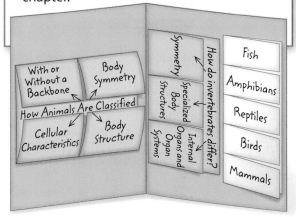

## Use Vocabulary

*Write the vocabulary term that best matches each phrase.*

**1** body plan that can be divided into two nearly equal parts anywhere through its central axis

**2** body plan that cannot be divided into two nearly equal parts

**3** structure that develops into a backbone in vertebrates

**4** a structure such as a leg or an arm

**5** two sides that are nearly mirror images of each other

**6** a thick, hard covering on arthropods

## Link Vocabulary and Key Concepts

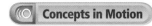 Concepts in Motion   Interactive Concept Map

*Copy this concept map, and then use vocabulary terms from the previous page to complete the concept map.*

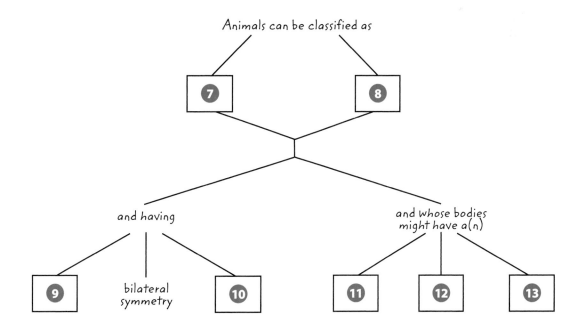

## Understand Key Concepts 🗝️

**1** Which characteristic does NOT apply to animals?

A. collagen
B. photosynthesis
C. digestive system
D. eukaryotic cell

**2** What characteristic applies to the animal shown below?

A. asymmetry
B. vertebrate
C. bilateral symmetry
D. radial symmetry

**3** Which characteristic separates the animal kingdom into two categories?

A. backbone
B. DNA
C. notochord
D. symmetry

**4** Which characteristics do all cnidarians have in common?

A. free swimming, radial symmetry
B. radial symmetry, attached to rocks
C. tentacles, bilateral symmetry
D. tentacles, stinging cells

**5** Which characteristic is NOT shared by both flatworms and nematodes?

A. bilateral symmetry
B. a simple brain
C. live in moist environments
D. two openings in digestive system

**6** To which phylum does the animal below belong?

A. Annelida
B. Mollusca
C. Nematoda
D. Platyhelminthes

*Use the figure below to answer questions 7 and 8.*

**7** What characteristic distinguishes the animal shown above from a fish?

A. bones
B. notochord
C. nerve cord
D. no gills

**8** To which phylum does this animal belong?

A. Annelida
B. Cnidaria
C. Chordata
D. Mollusca

# Chapter Review

## Critical Thinking

**9** **Create** a table that compares the three types of symmetry in the animal kingdom.

**10** **Analyze** Why is digestion important to animals, but not to plants?

**11** **Infer** Animals have cells specialized for different functions. Why is this feature an advantage for survival in a multicellular organism?

**12** **Analyze** Explain how you could determine whether the animals shown below are from the same phylum or different phyla.

**13** **Evaluate** Why are sponges, cnidarians, flatworms, and roundworms limited to life in water or moist environments?

**14** **Compare** What characteristics do fish and lancelets have in common? How are they different?

**15** **Infer** Why do most vertebrates have appendages such as fins, wings, and legs, whereas most invertebrates do not?

## Writing in Science

**16** **Write** a paragraph giving two ways a scuba diver could tell the difference between a sea anemone and a sea slug during an ocean dive.

## REVIEW THE BIG IDEA

**17** In what ways does a body with an internal skeleton have an advantage over a body with no internal or external support?

**18** What are the major groups of animals, and how do they differ?

## Math Skills ✕⁄₊ 

— Review
Math Practice —

### Use Percentages

**19** Worldwide, there are about 300,000 species of Lepidoptera, of which an estimated 14,500 are butterflies. What percentage of Lepidoptera species are butterflies?

**20** Of the estimated 1.2 million species of invertebrates, about 40,000 are crustaceans. What percentage of invertebrates are crustaceans?

**21** Of the estimated 1.2 million species of invertebrates, about 950,000 are insects. What percentage of invertebrates are insects?

# Standardized Test Practice

*Record your answers on the answer sheet provided by your teacher or on a sheet of paper.*

## Multiple Choice

1 Which is a characteristic of all vertebrates?

  A digestive system with one opening

  B offspring hatch from eggs

  C respiratory system with lungs

  D spinal cord enclosed in bones

*Use the figure below to answer question 2.*

2 Which is true of the animal shown above?

  A It has bilateral symmetry.

  B It has radial symmetry.

  C It is an invertebrate.

  D It is a vertebrate.

3 Which animals make up Phylum Chordata?

  A insects, spiders, and crabs

  B mussels, octopuses, and squids

  C snails, slugs, and clams

  D vertebrates, lancelets, and tunicates

4 Which is a characteristic of all invertebrates?

  A backbone absent

  B exoskeleton present

  C symmetry absent

  D tube feet present

5 Which is NOT a characteristic of all animals?

  A digesting food

  B eating other organisms

  C having specialized cells

  D moving around

*Use the figure below to answer question 6.*

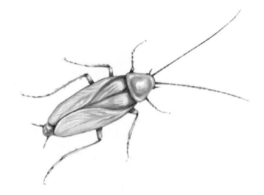

6 The animal in the figure above is classified in which phylum?

  A Arthropoda

  B Chordata

  C Echinodermata

  D Mollusca

7 Which animal has radial symmetry?

  A flatworm

  B jellyfish

  C octopus

  D sponge

*Use the figure below to answer questions 8 and 9.*

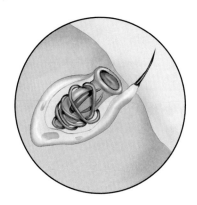

**8** Which animals contain the structure shown in the figure?

   **A** bees and wasps

   **B** jellyfish and corals

   **C** octopuses and squids

   **D** sea stars and sea urchins

**9** How do animals use the structure shown in the figure?

   **A** for breathing

   **B** for feeding

   **C** for mating

   **D** for seeing

**10** Which class of vertebrates feeds milk to its young?

   **A** amphibians

   **B** fish

   **C** mammals

   **D** reptiles

## Constructed Response

*Use the figure below to answer questions 11 and 12.*

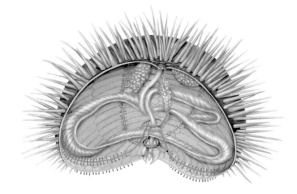

**11** Identify the phylum of the animal shown in the figure. What type of animal is it? Give examples of other animals in this phylum.

**12** Use the figure above to describe at least three characteristics typical of the phylum represented.

**13** Which adaptations enable birds and reptiles to live on land while fish and amphibians must live in or near water for at least part of their life cycles?

**14** List the four characteristics of all chordates. Explain the relationship of each characteristic to an adult human.

| NEED EXTRA HELP? | | | | | | | | | | | | | | |
|---|---|---|---|---|---|---|---|---|---|---|---|---|---|---|
| If You Missed Question... | 1 | 2 | 3 | 4 | 5 | 6 | 7 | 8 | 9 | 10 | 11 | 12 | 13 | 14 |
| Go to Lesson... | 3 | 1 | 1 | 2 | 1 | 1 | 1 | 2 | 2 | 3 | 2 | 2 | 3 | 3 |

# Animal Structure and Function

## THE BIG IDEA

Why do animals have different structures that perform similar functions?

### Inquiry  Feathers for Hearing?

The feathers around this great gray owl's eyes enable it to hear better. The feathers form discs around the eyes that funnel sounds toward the owl's ears.

- What other functions do you think feathers have for an owl?
- What other structures might enable the owl to survive?
- Why do animals have different structures that perform similar functions?

# Get Ready to Read

## What do you think?

Before you read, decide if you agree or disagree with each of these statements. As you read this chapter, see if you change your mind about any of the statements.

**1** All animals on Earth have internal skeletons.

**2** All animals that live in the water move using fins.

**3** Earthworms have an open circulatory system that transports blood and other substances throughout the body.

**4** Gills and lungs are different structures that perform the same function.

**5** The shape of an animal's teeth depends on its diet.

**6** Excretion is used only by animals that live on land.

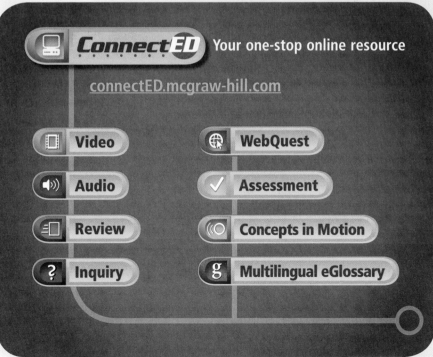

**ConnectED** Your one-stop online resource

connectED.mcgraw-hill.com

- Video
- Audio
- Review
- Inquiry
- WebQuest
- Assessment
- Concepts in Motion
- Multilingual eGlossary

## Reading Guide

### Key Concepts 🔑
**ESSENTIAL QUESTIONS**

- How are the types of support alike, and how are they different?
- How do the types of control compare and contrast?
- How do the types of movement compare and contrast?

### Vocabulary
**hydrostatic skeleton** p. 412

**coelom** p. 412

**nerve net** p. 414

**undulation** p. 416

g **Multilingual eGlossary**

🎞 **Video** **Science Video**

# Support, Control, and Movement

## Inquiry How does it move?

The animal shown above moves through its environment using a unique motion. The structures in its body enable it to move. How does an animal's movement depend on the environment it lives in? What structures in its body enable it to move?

## How does an earthworm move?

You move by using your muscles and skeleton. However, an earthworm does not have a skeleton. How is an earthworm able to move?

1 Read and complete a lab safety form.

2 Place a **paper towel** in the bottom of a **plastic container.** Add water to the paper towel until it is damp, but not dripping wet.

3 Place an **earthworm** on the surface of the paper towel and observe the earthworm for several minutes.

4 Pay particular attention to what happens when the earthworm moves. Note the changes in the earthworm's body and the motion that enables it to move.

### Think About This

1. How do you think the body of an earthworm has structure even though it has no skeleton?

2. Describe how the shape and length of the different segments of the earthworm change to cause it to move.

3. **Key Concept** How do you think the structure and movement of an earthworm is different from that of animals with skeletons?

## The Importance of Support, Control, and Movement

Think about the different environments where animals live. Some animals live their entire lives in water. Others live only on land. Regardless of their environments, all animals have the same basic needs: food, water, and oxygen. However, in order to survive in different habitats, animals have different structures with similar functions.

Fish and birds live in dramatically different environments. However, both use structures to obtain oxygen from their environments. In a similar way, animals have different structures for support, control, and movement. Without these, animals, such as the goats in **Figure 1,** could not obtain the things they need to survive. In this lesson you will read about how animals in different habitats use different structures to provide support and control for their bodies and to move around.

**Figure 1** These goats have pads between their hooves that enable them to climb trees and reach food.

## Structures for Support

As you have just read, organisms have structures to provide support, control, and movement. What structures provide support? Most animals are invertebrates, or animals without backbones. Animals with backbones, such as humans, are called vertebrates. Vertebrate and invertebrate animals have different types of structures that provide support.

### Hydrostatic Skeletons

Filling a balloon with water gives it shape. This is because the force of the water against the surface of the balloon gives the balloon structure. Just as the water in a balloon provides structure, many organisms use internal fluids to provide support. *A* **hydrostatic skeleton** *is a fluid-filled internal cavity surrounded by muscle tissue. The fluid-filled cavity is called the* **coelom** (SEE lum). Muscles that surround the coelom help some organisms move by pushing the fluid in different directions. Earthworms, such as the one shown in **Figure 2,** jellyfish, and sea anemones (uh NE muh neez) are organisms that have hydrostatic skeletons. Organisms with hydrostatic skeletons do not have bones or other hard structures that provide support.

 **Reading Check** What type of skeleton do jellyfish have?

### Hydrostatic Skeletons

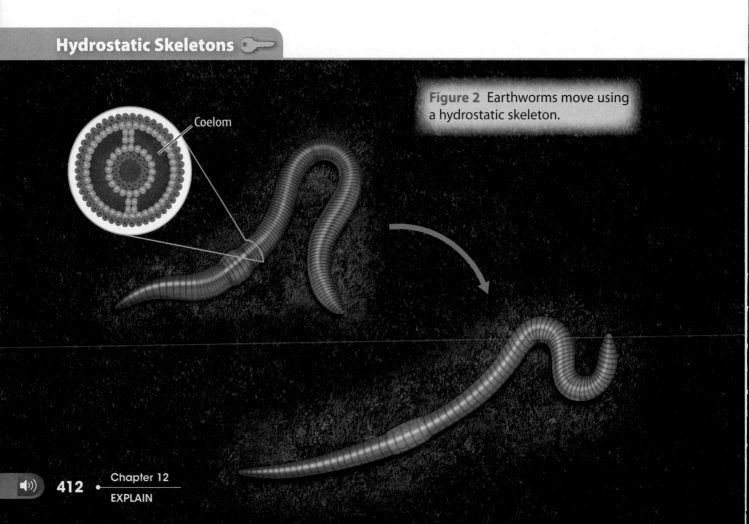

Coelom

Figure 2 Earthworms move using a hydrostatic skeleton.

## Exoskeletons

Some organisms get support from structures on the outside of the body. Hard outer coverings, called exoskeletons, provide support and protection for many invertebrates. A hard exoskeleton protects internal tissues from predators or damage. Exoskeletons are sometimes called shells in species such as crabs and snails. In some species, such as the lobster shown in **Figure 3,** the exoskeleton does not grow as the animal grows. It must be shed when it gets too small, leaving the animal defenseless until the new exoskeleton forms and hardens.

## Endoskeletons

Peaches have a soft, fleshy exterior that covers a hard seed. The bodies of many animals are similar in that they also have a covering over hard internal structures. These internal support structures in animals such as fish, birds, and mammals are called endoskeletons. Most endoskeletons, such as the one shown in **Figure 3,** are made of bone. Some, for example the endoskeletons of sharks, are made of cartilage. An endoskeleton protects internal organs and provides the organism with structure and support. Tortoises and turtles are unique because they have both endoskeletons and exoskeletons. The endoskeleton protects the organs and the hard exoskeleton shell protects the animal from predators.

 **Key Concept Check** How are the types of support alike, and how are they different?

**Figure 3** Most organisms get support from either exoskeletons or endoskeletons.

✅**Visual Check** How do the support structures of the squirrel and the lobster differ?

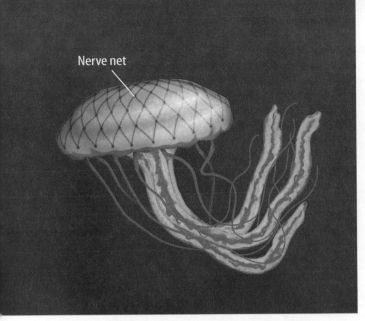

Nerve net

**Figure 4** 🔑 The nerve net of this jellyfish enables it to respond to its environment.

# Structures for Control

All animals react to changes in their environments. Just as different animals have different structures for support, they also have different control systems. These control systems, called nervous systems, help protect animals from harm and help animals move and find food.

## Nerve Nets

Animals with radial symmetry and no brain have nerve nets with a central ring that control their bodies. *A* **nerve net** *is a netlike control system that sends signals to and from all parts of the body.* Signals sent through the nerve net and ring cause an organism's muscle cells to contract. These contractions help the animal move. Cnidarians (nih DAYR ee unz) such as jellyfish and sea anemones have nerve nets that sense physical contact and detect food. Nerve nets and rings help the jellyfish shown in **Figure 4** move and capture prey.

## Inquiry MiniLab

**10 minutes**

### How do nerve nets and nerve cords function? 

Animals use different types of systems to sense the environment and react. Animals that do not have a brain use nerve nets to send signals throughout their bodies. Animals with brains use a nerve cord.

1 Read and complete a lab safety form.

2 One of your classmates will use a **stopwatch** to time the procedure.

3 Use **segments of string** to connect to your classmates in the shape of a nerve net. When you have formed the structure, close your eyes and bow your head.

4 When you receive the signal, gently tug on the next student's nerve-cell string. When you feel a tug on your string, raise your head, open your eyes, and tug the ends of the string connected to the people next to you.

5 Form a nerve-cord structure and repeat the activity. Record the results of your activities in your Science Journal.

### Analyze and Conclude

1. **Assess** which nerve structure took longer to reach the final student.

2. 🔑 **Key Concept** Describe one advantage and one disadvantage of using each type of structure to convey information.

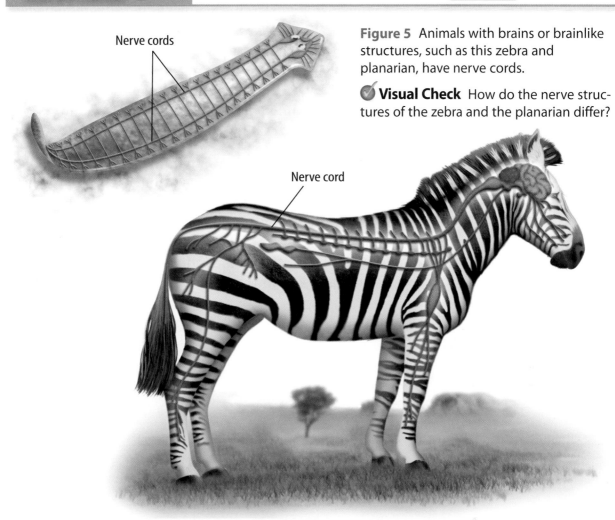

Nerve cords

Nerve cord

**Figure 5** Animals with brains or brainlike structures, such as this zebra and planarian, have nerve cords.

✔️ **Visual Check** How do the nerve structures of the zebra and the planarian differ?

## Nerve Cords

Animals with bilateral symmetry have brains or brainlike structures to detect and respond to their environments. An animal with a brain or a brainlike structure has a nerve cord, as shown in **Figure 5.** An animal with a nerve cord usually has many **neurons** that detect changes in its external environment. Signals detected by neurons are sent to the nerve cord, which might initiate a reflex response, and then to the brain for processing. Just as a telephone wire transmits signals between two buildings, the nerve cord enables signals to move between neurons and the brain. In vertebrates, nerve cords are also called spinal cords.

🔑 **Key Concept Check** How do the types of control compare and contrast?

**REVIEW VOCABULARY**

**neuron**
basic functioning unit of the nervous system

## Types of Movement

All animals move at some point in their lives. Some animals, such as birds or tigers, move around throughout most of their lives. Other animals, such as sponges, move during only part of their lives. Movement helps an animal obtain food and escape from danger. Because different animals live in different habitats, they use different structures to move around.

### Undulate Motion

It might be easy to figure out how an animal with legs moves around. But do you know how animals that do not have legs move? *Some animals move in a wavelike motion called* **undulation** (un juh LAY shun). Animals that move by undulation, such as snakes, fishes, and the eel in **Figure 6,** use their muscles to push their bodies forward. Undulation is used by animals that live on land and in the water.

**WORD ORIGIN** .............

**undulation**
from Late Latin *undulatus,*
means "wavy"

### Undulation 🔑

**Figure 6** Snakes, eels, and other animals move by undulating their bodies.

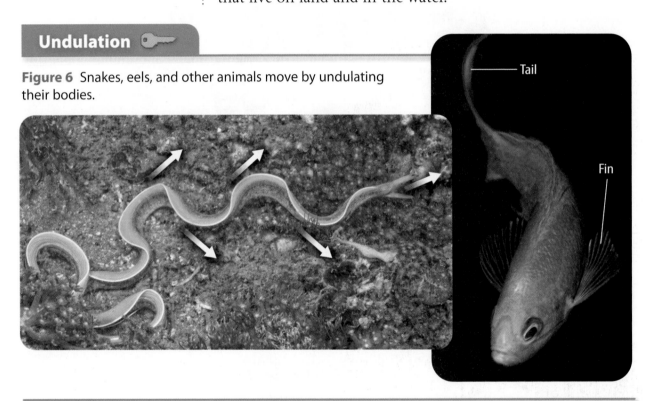

Tail

Fin

### Swimming

Many animals that live in water move by swimming. Some animals, such as the fish shown in **Figure 6,** use their fins and tails to move through water. Other animals, such as octopuses, take in water and then push the water out forcefully to move forward in a process called jet propulsion. You might already know that many organisms, such as humans and dogs, also can swim by moving their arms and legs, even though they do not live in water.

## Walking

Most animals that live on land move by walking. The body's weight rests on two, four, six, or eight legs and shifts when the legs move. Some animals, such as rabbits and frogs, also are capable of jumping using their limbs.

## Flying

Many animals move through the air by flying. Birds, some insects, and bats all use wings to move around. Wings, such as those shown in **Figure 7,** are a type of limb. By moving their wings, animals can lift their bodies and keep them in the air. Animals that have wings also have legs that are used to move around on land.

Wings are not the only structures that enable animals to move through the air. Some animals can glide or move through the air without flapping their limbs. Some species of fish have large fins that are used to glide short distances to escape predators. Some squirrels, marsupials, and even snakes can glide. They launch themselves from a high point and glide down by flattening their bodies or stretching out tissues to form a structure similar to a parachute.

 **Key Concept Check** How do the types of movement compare and contrast?

**Figure 7** Many birds move by flying, while a flying squirrel has the ability to glide.

# Lesson 1 Review

## Visual Summary

Animals have different structures for support, control, and movement.

Some animals have hydrostatic skeletons.

Most animals with wings move by flying.

**FOLDABLES®**

Use your lesson Foldable to review the lesson. Save your Foldable for the project at the end of the chapter.

## What do you think NOW?

You first read the statements below at the beginning of the chapter.

**1.** All animals on Earth have internal skeletons.

**2.** All animals that live in the water move using fins.

Did you change your mind about whether you agree or disagree with the statements? Rewrite any false statements to make them true.

## Use Vocabulary

**1** **Use the terms** *hydrostatic skeleton* and *coelom* in a sentence.

**2** **Define** *endoskeleton* in your own words.

**3** Eels contract their muscles and move forward by a process called _____.

## Understand Key Concepts

**4** Which is used by a jellyfish to sense and respond to changes in the environment?
  - **A.** coelom
  - **B.** undulation
  - **C.** nerve cord
  - **D.** nerve net

**5** **Explain** how a turtle's support system enables it to survive.

## Interpret Graphics

**6** **Analyze** how the control system shown below helps cnidarians respond to changes and capture prey.

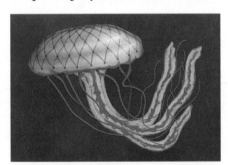

**7** **Organize Information** Copy and fill in the table below to describe how animals move in their habitats.

| Type of Movement | Type of Limb | Habitat |
|---|---|---|
| Undulation | | |
| Swimming | fins | |
| Walking | | land |
| Flying | | |

## Critical Thinking

**8** **Relate** a bird's limbs to its ability to walk and fly.

**9** **Assess** the role of the coelom in providing earthworms with support.

# Jet Propulsion

## The Secret of a Squid's Speed

A squid swims slowly along the ocean floor, flapping its delicate fins. Suddenly, it spots a shark approaching. In a flash, the squid darts away and is gone. When a squid has to move fast, its fins can't get the job done. It uses jet propulsion.

Think of what happens when you let go of a balloon as you're blowing it up. Air rushes out of the balloon in one direction, launching it in the opposite direction. This movement is an example of jet propulsion. Cephalopods (SE fuh luh podz) , animals such as squids, jellyfishes, and octopuses, use jet propulsion to move quickly through the ocean. However, they shoot water out of their bodies instead of air.

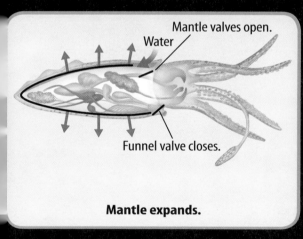

Water
Mantle valves open.
Funnel valve closes.
**Mantle expands.**

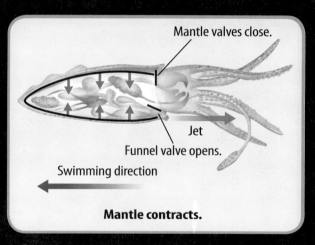

Mantle valves close.
Jet
Funnel valve opens.
Swimming direction
**Mantle contracts.**

Ⓐ A squid opens its mantle valves, drawing in water. Then the mantle valves close so the water can't escape.

Ⓑ The squid contracts its mantle, opens its funnel valve, and shoots out water through the funnel. This propels the squid through the water in the opposite direction. A squid can change directions by bending its funnel the other way.

AMERICAN MUSEUM OF NATURAL HISTORY

## It's Your Turn

**DIAGRAM** Work with a partner to research another animal that uses jet propulsion. Then draw and label a diagram that you can use to explain your findings to the class.

# Circulation and Gas Exchange

## Reading Guide

### Key Concepts 🔑
ESSENTIAL QUESTIONS

- How do the types of gas exchange differ?
- What are the differences between open and closed circulatory systems?

### Vocabulary

**diffusion** p. 422

**spiracle** p. 422

**gills** p. 423

**open circulatory system** p. 424

**closed circulatory system** p. 425

**g** Multilingual eGlossary

## Inquiry Underground Tunnels?

The large, orange structure might look like the entrance to an underground system of tunnels, but it is part of an insect! Tiny holes such as this enable some organisms to exchange gases directly with the environment.

### Which system is faster?

In some animals, blood surrounds organs. In other animals, blood is carried through vessels. Which system can transport oxygen and nutrients more efficiently?

1 Read and complete a lab safety form.

2 Fill a **large plastic bowl** with water. Center a **coin** on the bottom of the bowl as a target. Place three **marbles** around the inside bottom of the bowl.

3 Have your partner time you with a **stopwatch.** Use a **turkey baster** to move each of the marbles onto the target by pushing water behind each marble. Do not touch any of the marbles directly. Stop timing when all marbles have touched the target.

4 Remove water from the bowl until it is 1/3 full. Center the coin on the bottom of the bowl. Place the three marbles in a small **beaker** of water.

5 Take a length of **plastic tubing** and aim one end at the target in the water. Have your partner start timing you. Pour the marbles and water from the beaker into the other end of the tube so that the marbles flow through the tube and strike the target in the bowl. Stop timing when the third marble touches the target.

#### Think About This

1. Which system was able to deliver the marbles to the target faster? Why do you think one system is faster than the other?

2. What materials are represented by the marbles? What is represented by the target?

3. 🔑 **Key Concept** Which system do you think might work best for slow-moving animals? For fast-moving animals? Explain your reasoning.

## The Importance of Gas Exchange and Circulation

All cells need nutrients and oxygen to survive. Recall that animals obtain nutrients and oxygen from their environment. Organisms must take in these substances and get them to each cell. Structures in animal bodies transport the substances to all cells. They also help remove wastes such as carbon dioxide from the body. You might recall that most of an animal cell is made of water. In addition to nutrients, oxygen, and wastes, water also is transported throughout the body.

As with support, control, and movement, different animals use different structures to exchange gases and move substances throughout the body. The type of system used depends on the animal's habitat. In this lesson, you will read about the different structures that animals have that help cells exchange gases and get nutrients and oxygen.

**FOLDABLES**

Make a vertical book. Label it as shown. Use it to organize your notes about gas exchange and circulatory systems.

Gas Exchange Systems | Circulatory Systems

## MiniLab

**10 minutes**

### How do the surface areas of different respiratory systems compare?

The respiratory systems of animals need a large surface area to perform gas exchange. Which system has the most surface area?

1. Read and complete a lab safety form.

2. Use **paper** and **scissors** to create a model of gills and a model of book lungs. Calculate the surface areas of the completed models. Record the data in your Science Journal.

3. Create a model of the alveoli in a lung. Wrap paper around **marbles** to model alveoli. Use **rubber bands** or **string** to hold the paper around the marbles. Calculate the surface area of the model.

4. When your models are completed, unfold each of them and measure the surface area of each model. Record the data in your Science Journal.

### Analyze and Conclude

1. **Compare and contrast** the surface area of each model before and after you unfolded it. How does it differ?

2. **Analyze** how folding affects the structures of the respiratory system in terms of size and surface area.

3. 🔑 **Key Concept** Infer why the amount of surface area might be important in determining the respiratory rates of organisms.

## Gas Exchange

All animals must take in oxygen and eliminate carbon dioxide to survive. Oxygen must enter the body so the cells and tissues are able to use it for life processes. However, different animals use various structures to perform gas exchange.

### Diffusion

The basic process of gas exchange requires no structures at all and is called diffusion. **Diffusion** *is the movement of substances from an area of higher concentration to an area of lower concentration.* In simple animals such as sponges, whose bodies are only a few cell layers thick, no special gas exchange structures are needed. Diffusion occurs through all parts of the body. Oxygen passes directly into cells from the environment. In a similar manner, waste gases leave cells and enter the environment. Other animals use specialized structures in addition to diffusion.

✔ **Reading Check** What is diffusion?

### Spiracles

Some organisms exchange gases through the sides of their bodies. **Spiracles** *are tiny holes on the surface of an organism where oxygen enters the body and carbon dioxide leaves the body.* Insects such as beetles and arachnids such as spiders have spiracles. Although beetles and spiders both have spiracles, they use different tissues to transport oxygen throughout the body. Beetles have structures called tracheal (TRAY kee ul) tubes, and spiders have folded structures called book lungs to take in oxygen.

Tracheal tubes, such as the ones shown in **Figure 8,** are hoselike structures that branch into smaller tubes. Much as a river branches off into smaller streams, smaller branches of tracheal tubes help oxygen get to more places in the body. In contrast, book lungs are stacks of folded wall-like structures. Although tracheal tubes and book lungs look different, they are both used for gas exchange.

## Structures for Gas Exchange 🔑

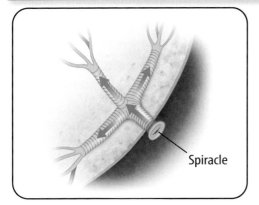

**Tracheal tubes**

Water flow

**Gills**

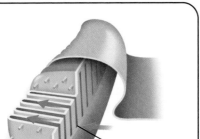

Spiracles

**Book lungs**

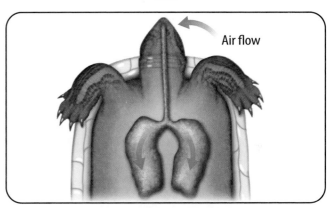

Air flow

**Lungs**

### Gills

Most animals that live in the water have gills for gas exchange. **Gills** *are organs that enable oxygen to diffuse into an animal's body and carbon dioxide to diffuse out.* In aquatic animals such as the fish shown in **Figure 8,** water enters the mouth to get to the gills. Oxygen in the water is taken in by gill filaments and then transported to the rest of the body. Gills also remove carbon dioxide from the body. Like other organs in the body, gills are surrounded by capillaries that help transport oxygen and carbon dioxide to and from cells.

### Lungs

Many animals that live on land, including the turtle in **Figure 8,** and some types of fish and snails have lungs for gas exchange. Lungs are baglike organs that can be filled with air. Once the lungs fill with air, oxygen diffuses into the capillaries within the lungs' tissues and carbon dioxide diffuses out of the animal's body. Recall that capillaries transport oxygen to other cells in the body through the circulatory system.

🔑 **Key Concept Check** How do the organs used for gas exchange differ?

**Figure 8** Spiracles, tracheal tubes, gills, and lungs are all used for gas exchange.

✓ **Visual Check** What structures for gas exchange involve spiracles?

**WORD ORIGIN** · · · · · · · · · · ·

**diffusion**
from Latin *diffundere*, means "to scatter"

# Circulation

You have just read about how gases are **exchanged** between animals and the environment. After an animal takes in oxygen, the oxygen has to travel to all parts of the body. Much like pipes in a house help transport water to the kitchen and bathrooms, an animal's circulatory system helps materials move through the body. Different animals have different circulatory systems. The type of circulatory system used often determines how quickly blood moves through the animal.

## Open Circulatory Systems

Snails, insects, and many other invertebrates have open circulatory systems. *An* **open circulatory system** *is a system that transports blood and other fluids into open spaces that surround organs in the body.* In an open circulatory system, such as the circulatory system of a bee shown in **Figure 9,** oxygen and nutrients in blood can enter all tissues and cells directly. Carbon dioxide and other wastes are taken up by blood surrounding the organs and removed from the body. Muscles help move blood through the body. It can take a long time for blood to move through an open circulatory system.

**Figure 9** Open circulatory systems transport blood into open spaces in the body. Closed circulatory systems transport blood through vessels.

## Open and Closed Circulatory Systems 🔑

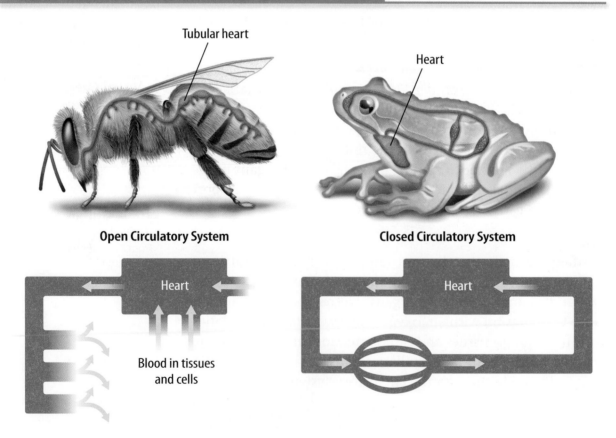

Tubular heart

Heart

**Open Circulatory System**

**Closed Circulatory System**

Heart

Blood in tissues
and cells

Heart

## Closed Circulatory Systems

Some animals, such as the tree frog shown in **Figure 9**, transport materials through another system called a closed circulatory system. *A **closed circulatory system** is a system that transports materials through blood using vessels.* Vessels help animals with closed circulatory systems move blood and other substances through the body faster than an open circulatory system.

As in an open circulatory system, muscles help blood move in a closed circulatory system. However, in a closed circulatory system, the muscles surround the blood vessels. These muscles contract and push blood through the vessels. They also can change the amount of blood flow. A closed circulatory system keeps plasma and red blood cells that carry oxygen separated from other fluids and structures in the body. Small blood vessels called capillaries surround organs and help oxygen and nutrients move from the circulatory system to cells in organs.

 **Key Concept Check** What are the differences between open and closed circulatory systems?

## Chambered Hearts

Different animals have hearts with different numbers of compartments called chambers, as shown in **Figure 10.** Fish have hearts with two chambers, whereas amphibian hearts consist of three chambers. Birds and mammals such as cats, dogs, and humans have hearts with four chambers. Almost all animals with three- or four-chambered hearts have lungs.

**Figure 10** Animals can have hearts with two, three, or four chambers.

**Visual Check** How would you describe an amphibian's circulatory system?

## Math Skills

### Use Proportions

A proportion is an equation with two ratios that are equivalent. Use proportions to solve problems such as the following: Veins hold about 55 percent of the body's blood. What is an organism's blood volume if the veins hold 2.6 L?

Set up the proportion.

$$\frac{55\%}{2.6\ \text{L}} = \frac{100\%}{x\ \text{L}}$$

Cross multiply.

$$55x = 260$$

Divide both sides by 55.

$$\frac{55x}{55} = \frac{260}{55} = 4.7\ \text{L}$$
$$x = 4.7\ \text{L}$$

### Practice

If a normal, complete heart cycle takes 0.8 s, how many cycles would the heart make in one day?

**Review**

- **Math Practice**
- **Personal Tutor**

 **Concepts in Motion** Animation

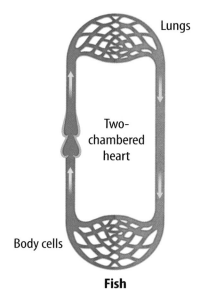

Lungs

Two-
chambered
heart

Body cells

**Fish**

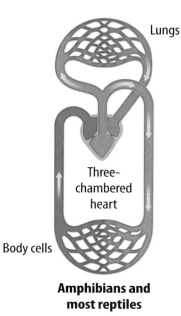

Lungs

Three-
chambered
heart

Body cells

**Amphibians and
most reptiles**

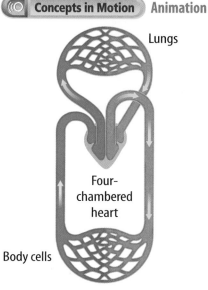

Lungs

Four-
chambered
heart

Body cells

**Crocodilians, birds
and mammals**

# Lesson 2 Review

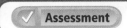 

## Visual Summary

Animals have different structures for gas exchange and circulation.

Animals can have open or closed circulatory systems.

Different animals have a different number of chambers in their hearts.

**FOLDABLES**

Use your lesson Foldable to review the lesson. Save your Foldable for the project at the end of the chapter.

## What do you think NOW?

You first read the statements below at the beginning of the chapter.

**3.** Earthworms have an open circulatory system that transports blood and other substances throughout the body.

**4.** Gills and lungs are different structures that perform the same function.

Did you change your mind about whether you agree or disagree with the statements? Rewrite any false statements to make them true.

## Use Vocabulary

1. **Use the term** *spiracles* in a sentence.

2. **Distinguish** between an open circulatory system and a closed circulatory system.

3. Aqatic animals use _____ to obtain oxygen from their environment.

## Understand Key Concepts

4. Which process helps oxygen move from the outside to the inside of cells?
   - A. absorption
   - B. circulation
   - C. diffusion
   - D. undulation

5. **Compare** the roles of book lungs and tracheal tubes in gas exchange.

6. **Infer** how the number of chambers in an animal's heart relates to its habitat.

## Interpret Graphics

7. **Identify** Copy and fill in the graphic organizer below with the ways animals exchange gases with their environment.

## Critical Thinking

8. **Hypothesize** how blood moves faster in a closed circulatory system when compared to an open circulatory system.

9. **Relate** the structures of gills and lungs to their roles in gas exchange.

## Math Skills

Review
— Math Practice —

10. In one experiment, the heart rate of a mollusk at rest was measured at 0.3 cycles/s. How many times did the mollusk's heart beat in 1 min?

# How do you determine what environment an animal lives in?

**Materials**

index cards

**Safety**

Animals have different structures and functions in order to survive in their environments. The combination of these different structures and functions makes animals unique from each other. How can you recognize the characteristics that determine the environments in which animals live and survive?

## Learn It

**Classifying** is the process of grouping objects or living organisms based on common features. Scientists classify animals according to the structures and functions they share.

## Try It

1. Label one blank index card for each of the characteristics listed in the categories below.

2. Shuffle the cards and spread them facedown on top of your desk.

3. Label four more index cards as follows: *Earthworm, Spider, Wolf,* and *Fish.* Shuffle these cards and place them facedown in a pile next to the cards you spread out.

4. Taking turns with your partner, choose one of the cards from the animal deck. Next, choose one of the cards spread out on the table and turn it over. If the characteristic you turned over applies to your animal, turn over another card and see if the second characteristic applies to your animal. If the characteristic does not apply to your animal, turn the card over, return it to the deck, and end your turn.

5. The object of the game is to select two cards in a row that apply to your animal. Each time you select a card, you must decide if it applies to your animal. If it does, select another card.

6. Score 1 point each time you can successfully select and turn over two cards that apply to the animal card you selected. Continue playing until all of the cards have been used.

## Apply It

7. **Record** the matches you made between characteristics and animals. Which matches did you complete?

8. **Describe** how some characteristics help animals compete better to find food or shelter.

9. 🔑 **Key Concept** Analyze the types of physical characteristics that make an animal best suited for a terrestrial or an aquatic environment. Explain how the different methods of gas exchange might be linked to these environments or other characteristics.

| Support Systems | Gas Exchange Systems | Circulatory Systems | Environments |
|---|---|---|---|
| Hydrostatic skeleton | book lungs | open circulatory system | aquatic environment |
| Exoskeleton | gills | closed circulatory system | terrestrial environment |
| Endoskeleton | lungs | | |
| | diffusion | | |

## Reading Guide

### Key Concepts
#### ESSENTIAL QUESTIONS

- How are an animal's structures for feeding and digestion related to its diet?

- How do the excretory structures of aquatic and terrestrial animals differ?

### Vocabulary

**crop** p. 432

**gizzard** p. 432

**absorption** p. 433

g  **Multilingual eGlossary**

# Digestion and Excretion

## Inquiry  What is it doing?

This caterpillar is chewing food—one step in the processes of digestion and excretion. Animals perform these processes to get the energy they need to live.

### What does it eat?

Humans and animals use several types of teeth to eat. Can you tell what an animal eats by looking at its teeth?

1 Incisors are teeth with a sharp edge in the shape of a wedge. Canines are pointy teeth. Molars have a large rough surface. Look at the photo of these teeth and answer the questions below.

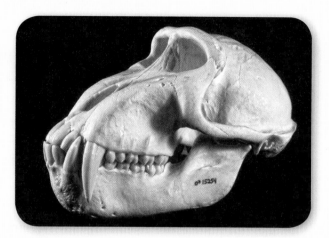

### Think About This

1. Which tooth do you think would be useful for cutting off the stem of a plant?

2. Which tooth might be helpful for tearing into flesh?

3. Which tooth might be used to grind up plants and meat?

4. 🔑 **Key Concept** Which type of teeth do you think is used by humans? Why?

## The Importance of Digestion and Excretion

You read in the first lesson that animals need nutrients to survive. Nutrients are obtained from food through digestion—the process of breaking down food into molecules that cells can absorb and use. After all nutrients are taken in, waste products not used by the body are removed by excretion. Excretion is important for survival because it removes harmful substances from the body. Just as different animals have different structures for gas exchange, they also have different structures to obtain and process nutrients and to remove wastes.

### Digestion

Animals have different structures for digestion, depending on what type of food they eat. For example, an animal that eats only seeds has a different set of structures from an animal that eats only meat. The first step of digestion usually happens when food is chewed, as shown in **Figure 11.** The food is further broken down by the digestive system in various ways, depending on the animal's diet. As you will read in this lesson, different animals use different structures to obtain and break down food.

✓ **Reading Check** How do animals obtain nutrients?

**Figure 11** The structures that this cow uses for digestion are good at breaking down grass and other plant matter.

## Structures for Feeding

For many animals, the first step in feeding is obtaining food. As with other functions of the body, animals use different structures to find and chew their food. You often can tell what type of diet an animal eats by looking at the structures that it has for feeding.

**Teeth** Many animals have teeth, one type of structure used for feeding. Different types of teeth are used to process different diets, as shown in **Figure 12.**

Animals that eat plants often have wide teeth used for chewing grass and other plants. Some have a few sharp teeth used to cut through twigs. Animals that eat insects have teeth with sharp points that are used for chewing.

Animals that eat only meat have several types of teeth. As shown in **Figure 12,** teeth in the front of the mouth are used for biting and holding food. Teeth in the rear of the mouth are pointed and used to cut up food. Animals that eat both plants and meat have sharp teeth that are used for cutting up food and wide, flat teeth that are used for grinding up food.

**Figure 12** 🔑 The shape of an animal's teeth depends on its diet.

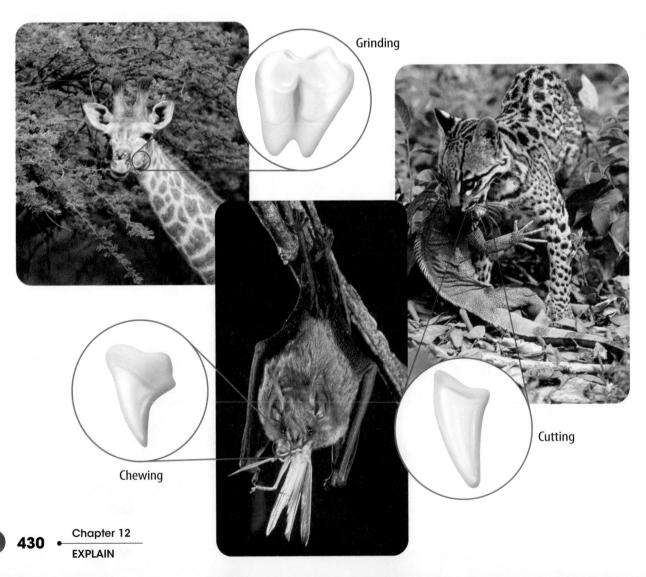

Grinding

Chewing

Cutting

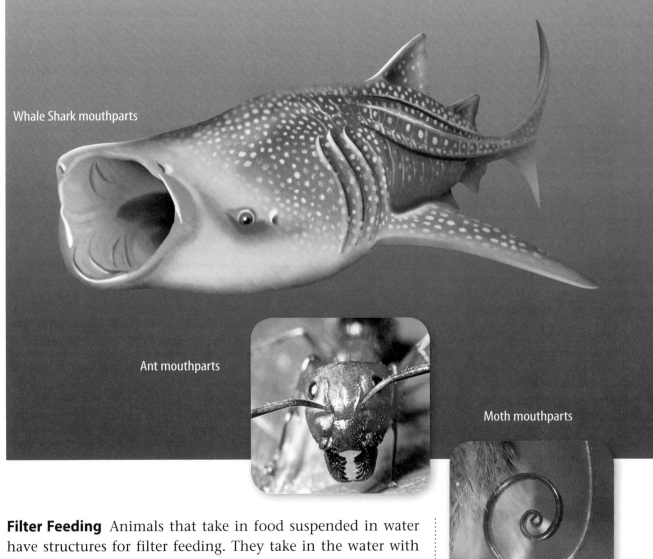

Whale Shark mouthparts

Ant mouthparts

Moth mouthparts

**Filter Feeding** Animals that take in food suspended in water have structures for filter feeding. They take in the water with the food, push the water out through a filtering structure, then eat the organisms that remain.

Some animals, such as certain whales, take a mouthful of water and push it out through baleen (bay LEEN). Baleen is a material similar to the bristles of a broom that filter out tiny organisms in the water. Certain types of sharks and fish filter food through their gills, such as the whale shark shown in **Figure 13.** Some animals, such as clams, filter feed without moving. They filter food from the water that moves around them. However, many filter feeders move around to find food. When flamingoes filter feed, they eat shrimp that is filtered through their beak from the water they take in.

**Mouthparts** Some animals, particularly insects, have specialized mouthparts for eating. Butterflies and moths use a long, tubelike mouthpart, shown in **Figure 13,** to get nectar from flowers. Ants and certain beetles have crushing jaws for ripping plant and animal matter.

**Key Concept Check** How are an animal's feeding structures related to its diet?

**Figure 13** ☞ Whale sharks and some insects have specialized mouthparts for feeding.

**Visual Check** How do mouthparts differ among the shark, the ant, and the moth?

## Structures for Digestion

After food is broken down into smaller parts by chewing, it breaks into even smaller components during digestion. Most animals have organs that form a specialized system for digestion. For example, many animals have stomachs and intestines that are used to digest food. The structures of the stomach and the intestine depend on the animal's diet. For example, animals such as cows and sheep that eat a lot of plant material have stomachs with several chambers. In each of these chambers, the tough plant material is processed so the animal can digest it.

**Crops** Some animals store their food in a crop before digesting it. *A* **crop** *is a specialized structure in the digestive system where ingested material is stored.* Many birds and insects have crops. Leeches, snails, and earthworms also have crops where they store undigested food. The crop in a leech can store blood and expands up to five times its body size.

**Gizzards** Animals without teeth that eat hard foods such as seeds sometimes have structures called gizzards. *A* **gizzard** *is a muscular pouch similar to a stomach that is used to grind food.* Some animals with gizzards, including certain birds, swallow rocks with their food. The rocks help break up the food.

SCIENCE USE V. COMMON USE

crop

*Science Use* a digestive system structure where material is stored

*Common Use* a plant or animal product that can be grown or harvested

### Inquiry MiniLab

**10 minutes**

#### How do gizzards help birds eat?

Some birds use gizzards to grind food into smaller pieces. Gizzards are small pouches that store food. The bird fills its gizzard with small stones that create the grinding action.

1. Read and complete a lab safety form.
2. Place 20 **sunflower seeds in the shell** in a **small, self-sealing plastic freezer bag.**
3. Fill another freezer bag about one-quarter full with **small stones.** Add 20 sunflower seeds in the shell.
4. Seal both bags and knead the contents of each bag for several minutes with your hands.
5. Open the bags and observe the condition of the seeds. Record your observations in your Science Journal.

#### Analyze and Conclude

1. **Compare and contrast** the condition of the sunflower seeds in the bag with stones to the sunflower seeds in the bag without stones.

2. **Analyze** how you think having a gizzard could be an advantage to a bird.

3. **Key Concept** Infer how the structure and function of the gizzard relates to the type of food that a bird can eat.

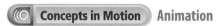

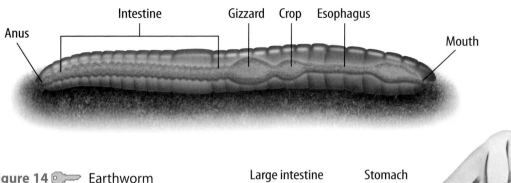

Anus    Intestine    Gizzard  Crop  Esophagus    Mouth

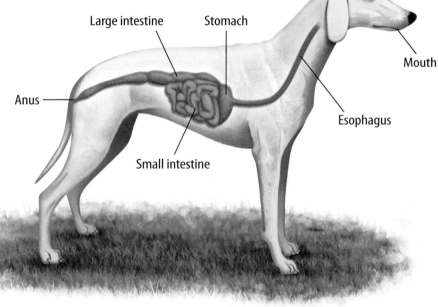

Large intestine    Stomach

Mouth

Anus

Esophagus

Small intestine

**Figure 14** 🔑 Earthworm digestive systems include a crop. Dog digestive systems include large and small intestines.

✅ **Visual Check** Why might the digestive systems of dogs and earthworms include different structures?

## Absorption

Whether an animal stores food before digestion or not, it must take the nutrients into the body in order to use them. **Absorption** *is the process in which nutrients from digested food are taken into the body.* Absorption happens as food moves through the digestive system, such as the ones shown in **Figure 14.** Many animals have digestive systems that contain enzymes. Enzymes are chemicals that help break food into small parts so that cells can absorb the nutrients.

In addition to enzymes, many organisms have structures that enable absorption. For example, many animals absorb nutrients in the intestine. After absorption, the structures used to move the nutrients throughout the animal's body also differ. In animals with a closed circulatory system, the capillaries that surround the intestine transport nutrients throughout the body. Recall that blood surrounds the organs of animals with open circulatory systems. In this case, nutrients enter the blood directly after absorption.

 **Key Concept Check** How are an animal's structures for digestion related to its diet?

**WORD ORIGIN** ·············

**absorption**
from Latin *absorbere*, means "to swallow up"

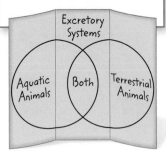

**FOLDABLES**

Make a vertical tri-fold Venn book. Label it as shown. Use it to compare and contrast the excretion process in animals from different habitats.

# Excretion

During the process of excretion, waste materials are removed from the body. Different animals excrete different types of wastes. The types of wastes animals excrete depend on the environments where they live.

## Diffusion

Recall that gas exchange occurs due to diffusion. While diffusion can bring in oxygen, it also can release carbon dioxide. Some organisms, such as sponges, have no filtering mechanisms in their bodies. Rather, waste materials are excreted as water moves in and out of the animal's pores.

## Excretion in Aquatic Animals

Many animals that live in aquatic environments, such as the fish shown in **Figure 15,** remove liquid wastes using kidneys. Most of the waste removed by the kidneys is water. The kidneys of fish also excrete other wastes, such as ammonia. Carbon dioxide is removed through the gills. Solid waste leaves the body in the form of feces.

## Excretion in Terrestrial Animals

Like aquatic animals, terrestrial animals also have kidneys. However, they excrete less water when removing wastes. Instead of excreting ammonia, most animals that live on land excrete urea as a waste product. Birds also excrete wastes, but they conserve water by excreting uric acid instead of ammonia or urea. Land animals excrete carbon dioxide through the lungs. They also excrete solid waste as feces.

 **Key Concept Check** How do the excretory structures of aquatic and terrestrial animals differ?

**Figure 15** Aquatic animals, such as this fish, use kidneys and gills to excrete wastes.

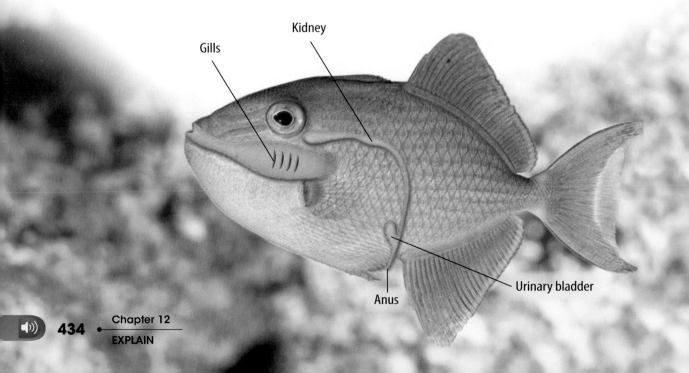

Gills

Kidney

Anus

Urinary bladder

# Lesson 3 Review

## Visual Summary

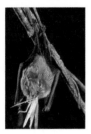

Animals have different structures for digestion and excretion.

Some organisms have structures that enable them to store food.

The type of waste an animal excretes depends on its environment.

**FOLDABLES**

Use your lesson Foldable to review the lesson. Save your Foldable for the project at the end of the chapter.

## What do you think NOW?

You first read the statements below at the beginning of the chapter.

**5.** The shape of an animal's teeth depends on its diet.

**6.** Excretion is used only by animals that live on land.

Did you change your mind about whether you agree or disagree with the statements? Rewrite any false statements to make them true.

## Use Vocabulary

**1** Nutrients are taken into the body by the process of _____.

**2** **Define** *gizzard* in your own words.

**3** Flamingoes and leeches have storage compartments for food called _____.

## Understand Key Concepts

**4** Which process moves nutrients from the digestive system into the circulatory system?

A. absorption       C. excretion
B. diffusion        D. undulation

**5** **Compare** the roles of gills and lungs in excreting carbon dioxide.

**6** **Explain** the role of gizzards in digestion.

## Interpret Graphics

**7** **Explain** the function of the system of structures shown below.

**8** **Summarize** Copy and fill in the graphic organizer below with the names of materials that animals excrete.

Wastes

## Critical Thinking

**9** **Relate** the shape of an animal's teeth to its diet.

**10** **Assess** the role of kidneys in aquatic and terrestrial animals.

## Materials

craft materials

## Safety

# Design an Alien Animal

You have read that animals have a variety of structures and functions in order to survive. The type of structures that a particular animal has depends on the environment where it lives. Your task is to imagine a unique environment on another planet. Be creative and describe the alien planet in detail. Then use what you know about animal structures and functions to imagine what life-forms live on the planet. The life-forms you design must be able to survive in the unique environmental conditions of the planet you described.

## Question

How do specialized structures and abilities help animals survive in specific environments? What animal characteristics would you expect to find in different environments?

## Procedure

1  Read and complete a lab safety form.

2  In your Science Journal, write a description of an imaginary alien planet. Be creative in describing your planet. Write down as many details as you can think of, including:

- Is the planet dry, or covered with water?

- What temperatures are experienced on your planet?

- What sources of food are present?

3  Include drawings along with your descriptions. Use labels to add detailed information to the environments you create.

4  Think about how the different animal species you studied have changed over time to live in different environments. Write a statement about the relationship between a species' characteristics and its environment. Predict how the environments you have created will affect organisms on your planet.

| Alien Structure and Function ||
|---|---|
| Structure | Function in Environment |
| Fins | |
| Claws | |
| | |
| | |

**5** Design several animals that will inhabit your alien planet. Create a data card for each animal that includes a picture of the animal. Add a chart like the one on the previous page to each data card that describes the structures of the animal and the functions of the structures. Include detailed drawings and/or descriptions of the body systems of each animal.

**6** Use the information you've gathered to create a three-dimensional model, or diorama, that illustrates your environment and the animals you have described.

## Analyze and Conclude

**7** **Explain** For what environmental conditions has your organism developed structures?

**8** **Analyze** How have some organisms gained an advantage over other organisms on the planet?

**9** **Compare and Contrast** How are the animals on your planet similar to and different from each other in terms of control, support, movement, gas exchange, feeding, digestion, and excretion? How do these characteristics help them maintain homeostasis?

**10** **The Big Idea** Why do the animals on your alien planet have different structures, even though they all live on the same planet?

## Communicate Your Results

Share your diorama with the rest of the class. Compare your planet's environment and life-forms with the ones your classmates created. What similarities and differences do you notice?

 **Extension**

Pair up with another student. Choose an organism from your partner's planet to bring to your planet. Write a prediction of what would happen. Would the new organism survive? How would the species adapt to the new environment? What organisms would it compete with? Would it threaten the survival of other organisms upon its arrival?

**Lab Tips**

☑ Consider your animal's environment when determining its diet.

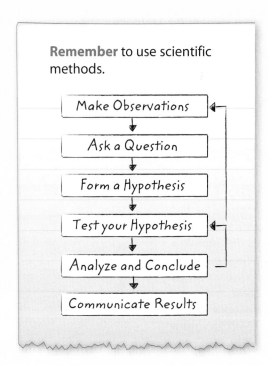

**Remember** to use scientific methods.

Make Observations
↓
Ask a Question
↓
Form a Hypothesis
↓
Test your Hypothesis
↓
Analyze and Conclude
↓
Communicate Results

# Chapter 12 Study Guide

 **Animals have different structures that perform similar functions which enable them to survive in different environments.**

| Key Concepts Summary  | Vocabulary |
|---|---|
| **Lesson 1: Support, Control, and Movement** <br><br> • Support structures give internal organs protection. Endoskeletons are made of bone or cartilage. Exoskeletons form shells made of minerals. <br><br> • **Nerve nets** detect changes in the environment over a large area without a brain. Nerve cords sense the environment, send the information to the brain for processing, and transmit the response to neurons. <br><br> • Animals have different structures that enable them to move through their habitats. | **hydrostatic skeleton** p. 412 <br><br> **coelum** p. 412 <br><br> **nerve net** p. 414 <br><br> **undulation** p. 416 |
| **Lesson 2: Circulation and Gas Exchange** <br><br> • Gas exchange occurs through **gills** in aquatic animals and through lungs or **spiracles** in terrestrial animals. <br><br> • In **open circulatory systems**, blood surrounds all organs in the body. **Closed circulatory systems** use blood vessels to transport substances throughout the body.    | **diffusion** p. 422 <br><br> **spiracle** p. 422 <br><br> **gills** p. 423 <br><br> **open circulatory system** p. 424 <br><br> **closed circulatory system** p. 425 |
| **Lesson 3: Digestion and Excretion** <br><br> • An animal's feeding structures depend on its diet. Animals that eat meat have sharp teeth that cut and tear, and animals that eat plants have wide, flat teeth for grinding.  <br><br> • Aquatic animals use kidneys to excrete large amounts of water and ammonia. Terrestrial animals use kidneys to excrete smaller amounts of water and either urea or uric acid. Aquatic animals excrete carbon dioxide through gills, and terrestrial animals excrete carbon dioxide through the lungs or spiracles. | **crop** p. 432 <br><br> **gizzard** p. 432 <br><br> **absorption** p. 433 |

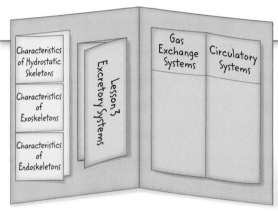
**FOLDABLES** **Chapter Project**

Assemble your lesson Foldables as shown to make a Chapter Project. Use the project to review what you have learned in this chapter.

Characteristics of Hydrostatic Skeletons

Characteristics of Exoskeletons

Characteristics of Endoskeletons

Lesson 3 Excretory Systems

Gas Exchange Systems

Circulatory Systems

## Use Vocabulary

1. Use the term *coelom* in a sentence.

2. Some animals store food in structures called _____.

3. Tiny openings on the surface of some animals that are used to take in oxygen are _____.

4. Define the term *crop* in your own words.

5. Snakes and eels move by _____.

6. Define the term *diffusion* in your own words.

## Link Vocabulary and Key Concepts

**Concepts in Motion** Interactive Concept Map

*Copy this concept map, and then use vocabulary terms from the previous page to complete the concept map.*

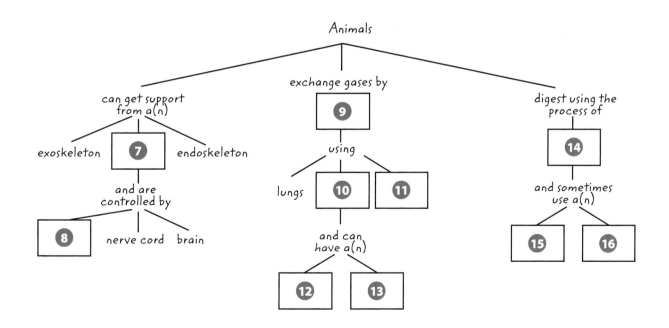

## Understand Key Concepts

**1** Which is NOT used to provide an animal with structural support?

A. endoskeleton
B. exoskeleton
C. hydrostatic skeleton
D. nerve net

**2** The fluid-filled sac in an earthworm is called a(n)

A. coelom.
B. endoskeleton.
C. nerve cord.
D. nerve net.

**3** The animal pictured below has which type of support system?

A. endoskeleton
B. exoskeleton
C. hydrostatic skeleton
D. no skeleton

**4** Animals with _____ do not have brains.

A. endoskeletons
B. gills
C. nerve cords
D. nerve nets

**5** Which structure is used for food storage?

A. crop
B. gizzard
C. kidney
D. stomach

**6** Capillaries are found in animals with

A. coeloms.
B. spiracles.
C. closed circulatory systems.
D. open circulatory systems.

**7** Which is NOT used for gas exchange?

A. crops
B. gills
C. book lungs
D. tracheal tubes

**8** The animal below uses which organs to excrete ammonia?

A. gills
B. intestines
C. kidneys
D. lungs

**9** Which is NOT used by aquatic animals to move through the water?

A. fin
B. gill
C. tail
D. wing

**10** What is the basic process of gas exchange?

A. absorption
B. circulation
C. diffusion
D. excretion

## Critical Thinking

**11 Describe** how the structure of cnidarians helps them respond to stimuli from all directions.

**12 Compare** the roles of endoskeletons and exoskeletons in providing animals with protection.

**13 Assess** the role of undulation in helping animals without appendages move.

**14 Relate** the structure of an animal's circulatory system to the rate at which blood moves throughout its body.

**15 Relate** the structure an organism uses for gas exchange to its habitat.

**16 Describe** how diffusion helps animals exchange gases.

**17 Relate** an animal's habitat to the amount of water it excretes.

**18 Hypothesize** how the structure pictured below helps an animal obtain nutrients without having to eat more food.

**19 Compare** the structure of teeth in animals that eat plants and animals that eat other animals.

*Writing in Science*

**20 Write** a five-sentence paragraph that describes how spiracles work together with book lungs and tracheal tubes to obtain oxygen. Be sure to include a topic sentence and a concluding sentence in your paragraph.

**REVIEW** THE B**I**G IDEA

**21** Why do animals have different structures that perform similar functions? Compare and contrast some structures that appear different but perform the same function.

**22** The owl shown below has various structures that enable it to survive. Compare these structures with an animal that lives in a different environment. List the similarities and differences.

**Math Skills**

Review
— Math Practice —

### Use Proportions

**23** Birds and mammals have four-chambered hearts. Out of a sample with 52,750 species of animals, 9,600 are birds and 5,500 are mammals. What percentage of animals in the sample have four-chambered hearts?

**24** The average human heart beats about 3 billion times during a lifetime. If the average heart rate is 70 beats/min, what is the average life span, in years, for the human heart? [Hint: 525,600 min = 1 y]

*Record your answers on the answer sheet provided by your teacher or on a sheet of paper.*

## Multiple Choice

**1** Which term describes a protective support structure found on the outside of certain organisms?

   **A** endoskeleton

   **B** exoskeleton

   **C** cartilage skeleton

   **D** hydrostatic skeleton

*Use the figure below to answer question 2.*

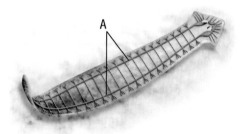

**2** What parts are labeled with the letter *A* in the figure?

   **A** bones

   **B** intestines

   **C** nerve cords

   **D** nerve nets

**3** How do the excretory structures of fish compare with those of birds?

   **A** Birds excrete feces and fish do not.

   **B** Birds have gizzards and fish have crops.

   **C** Fish have gills and birds have lungs.

   **D** Fish have kidneys and birds do not.

**4** Which animal moves by undulation?

   **A** a seagull

   **B** a snake

   **C** a squid

   **D** a squirrel

**5** What is the function of a crop?

   **A** exchange gases

   **B** protect organs

   **C** store food

   **D** transport material

*Use the figure below to answer question 6.*

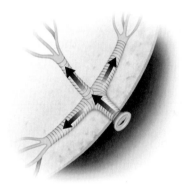

**6** Which gas exchange structure is shown?

   **A** a gill

   **B** a lung

   **C** a book lung

   **D** a tracheal tube

**7** How do terrestrial animals excrete carbon dioxide?

   **A** through gills

   **B** through inhalation

   **C** through lungs

   **D** through undulation

**8** Which animal has a nerve net that controls its body?

   **A** a dog

   **B** a python

   **C** a flying squirrel

   **D** a sea anemone

*Use the figure below to answer question 9.*

**9** A biologist in the field finds a large animal tooth like the one shown. Which describes how a tooth of this shape and size was most likely used?

  **A** to bite into large prey

  **B** to chew insects

  **C** to eat grass

  **D** to eat leaves from trees

**10** What gives a hydrostatic skeleton support?

  **A** bones

  **B** cartilage

  **C** coelom

  **D** shell

## Constructed Response

*Use the diagrams below to answer questions 11 and 12.*

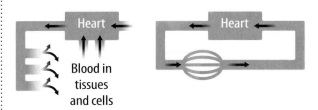

**11** Which type of circulatory system would be more efficient in moving blood through the body? Explain your answer.

**12** An open circulatory system allows oxygen and nutrients to enter all tissues and cells directly. A closed circulatory system allows blood to be moved through the body faster. Which system would enable an animal to process energy faster? Explain your answer.

**13** Explain how a book lung differs from the lung of a turtle.

**14** Give an example of an animal that has the ability to move by two different means.

| NEED EXTRA HELP? | | | | | | | | | | | | | | |
|---|---|---|---|---|---|---|---|---|---|---|---|---|---|---|
| If You Missed Question... | 1 | 2 | 3 | 4 | 5 | 6 | 7 | 8 | 9 | 10 | 11 | 12 | 13 | 14 |
| Go to Lesson... | 1 | 1 | 3 | 1 | 3 | 2 | 3 | 1 | 3 | 1 | 2 | 2 | 2 | 1 |

# Animal Behavior and Reproduction

**THE BIG IDEA**

How do animals communicate, interact, and reproduce?

**Inquiry** **What are they doing?**

This female chimpanzee is showing its offspring how to use a stick to find termites.

- How might this behavior be beneficial to the young chimpanzee?
- What other types of behaviors do animals have?
- In what ways might animals communicate, interact, and find mates?

# Get Ready to Read

## What do you think?

Before you read, decide if you agree or disagree with each of these statements. As you read this chapter, see if you change your mind about any of the statements.

**1** Animals react to their environments.

**2** All animal behavior is instinctive.

**3** Some animals give off light to communicate with each other.

**4** Animals always fight to protect their territories.

**5** During sexual reproduction, a sperm cell and an egg cell join.

**6** Some animals develop inside the mother.

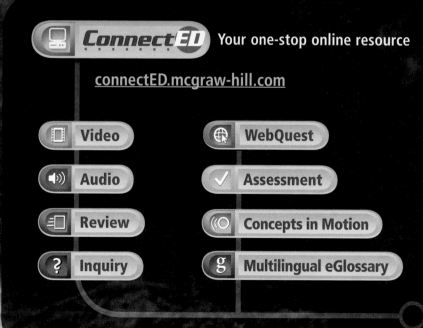

**ConnectED** Your one-stop online resource

connectED.mcgraw-hill.com

- Video
- WebQuest
- Audio
- Assessment
- Review
- Concepts in Motion
- Inquiry
- Multilingual eGlossary

# Types of Behavior

## Reading Guide

### Key Concepts 🔑
**ESSENTIAL QUESTIONS**

- How do behaviors help animals maintain homeostasis?
- How are animal behaviors classified?

### Vocabulary

**behavior** p. 447

**innate behavior** p. 449

**instinct** p. 450

**migration** p. 451

**hibernation** p. 451

**imprinting** p. 452

**conditioning** p. 453

🅖 **Multilingual eGlossary**

🎞 **Video** **BrainPOP®**

## Inquiry Sleeping?

This dormouse appears to be sleeping, but it is actually in a state of inactivity called hibernation. A dormouse hibernates during cold weather to conserve energy while food is scarce. Do you think the dormouse learned or was born knowing to hibernate during cold weather? What other behaviors might a dormouse exhibit?

# inquiry Launch Lab

**15 minutes**

## What happens when you touch a pill bug?

Pill bugs are arthropods that live under leaf litter and rocks. They have a special behavior that helps them defend themselves against other animals that might eat them.

1. Read and complete a lab safety form.

2. Obtain a **pill bug** and gently place it in a **petri dish** for observation. Study the pill bug without touching it, and draw it in your Science Journal.

3. Use a **cotton swab** to gently touch the pill bug, and observe it again. Draw the pill bug's reaction.

**Think About This**

1. How did the pill bug react when you touched its back?

2. What stimulus did you provide that was different from the pill bug's natural environment?

3. 🔑 **Key Concept** What other stimuli do you think might affect the pill bug?

## What is a behavior?

Have you ever watched a dog sniff the ground while it was out for a walk? Or have you seen a dog, such as the one in **Figure 1**, working with law enforcement and sniffing luggage at an airport? Why does a dog do this? Dogs receive information about their surroundings by sniffing. Dogs have a much more developed sense of smell than humans do. A dog's nose has about 220 million scent receptors, but a human's nose has only about 5 million.

The act of sniffing is a common dog behavior. *A* **behavior** *is the way an organism reacts to other organisms or to its environment.* Behaviors might be carried out by individual animals, such as a dog sniffing, or by groups of animals of the same species, such as a flock of birds flying together. Recall that organisms' bodies work to maintain a steady internal state called **homeostasis.** Behaviors are a way to maintain homeostasis when the environment changes.

🔑 **Key Concept Check** How do behaviors help animals maintain homeostasis?

**Figure 1** A dog's sniffing behavior helps it get information about its surroundings.

**REVIEW VOCABULARY** · · · · · · · · · ·

**homeostasis**
an organism's ability to maintain steady internal conditions when outside conditions change

When an animal carries out a behavior, it is reacting to a **stimulus** (STIHM yuh lus; plural, stimuli), or change. A stimulus can be external, such as the weather getting warmer, or internal, such as hunger. Scents coming from the pavement or a tree are external stimuli for a dog. The dog's response to the stimuli is sniffing.

### Stimuli

Stimuli can come in many forms and result in different behaviors. Changes in the external environment, such as a temperature change or a rainstorm, can affect an animal's behavior. Hunger, thirst, illness, and other changes in an animal's internal environment are stimuli, too.

### Responses to Change

Animals respond to changes and maintain homeostasis in different ways. For example, when the weather gets cooler, an organism might respond with a specific behavior. Birds, which must keep their bodies at the same temperature year-round, fluff their feathers and retain more thermal energy, as shown in **Figure 2.** The cooler weather is the stimulus, and the bird's feather fluffing is a response.

Animals also respond to internal stimuli, such as illnesses. If an animal is sick, its body might respond with a fever. The fever increases body temperature and might help the animal fight a disease. Vomiting is another response to an internal stimulus. A dog that ate something from the garbage might vomit to get the material out of its body. This behavior helps the dog maintain homeostasis by removing something that could cause an illness.

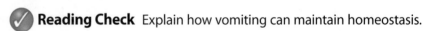 **Reading Check** Explain how vomiting can maintain homeostasis.

**Figure 2** 🔑 During warm weather, a bird's feathers are close to its body. When a bird fluffs its feathers during cold weather, it traps a layer of air around its body. The air helps keep the bird warm.

## Stress

Have you ever seen an animal run away when a human got too close? The human caused the animal to become stressed, and the animal reacted by running away. Some animals, such as the antelope shown in **Figure 3,** will almost always run away if they feel threatened. When an animal identifies a danger, its body prepares to either fight or run away from the perceived threat. This behavior is called the fight-or-flight response.

Not all animals have the same reaction and run away from dangerous situations. A wild male horse might attack another male in the same area to protect its herd. Some animals, such as rats, will run from danger but will fight if cornered.

## Innate Behaviors

As you have read, behaviors are responses to some type of stimulus. An animal's behaviors are a combination of those that are learned and those that are inherited and not linked to past experiences. *A behavior that is inherited rather than learned is called an* **innate behavior.**

An innate behavior happens automatically the first time an animal responds to a certain stimulus. For example, when tadpoles hatch, they already know how to swim. They do not learn how to do so by watching other tadpoles. Tadpoles can swim away from danger and find food as soon as they hatch.

Animals with short life spans have mostly innate behaviors. Animals such as insects rely on behaviors that they do not have to learn. They are able to find food and mates and avoid danger early in their lives. Insect innate behaviors include a cricket's ability to chirp and a moth's attraction to light. These types of behaviors enable animals to survive without learning from another animal.

**Figure 3** Some animals, such as antelope, respond to threatening situations by running away.

## Reflexes

Have you ever noticed what happens to the pupils in your eyes when you go into a dimly lit room? After a short period of time, they get larger. This happens without you thinking about it. This is an example of the simplest type of innate behavior, called a reflex. A reflex is an automatic response that does not involve a message from the brain.

Animals have reflexes, too. For example, an armadillo will jump straight upward about 1 m when startled, as shown in **Figure 4.** By jumping, the armadillo might be able to startle predators and escape.

## Instincts

Reflexes happen quickly and involve one behavior. Some innate behaviors involve a number of steps performed in a specific order. *A complex pattern of innate behaviors is called an* **instinct** (IHN stingt). Finding food, running away from danger, and grooming are some behaviors that are instincts in many animals.

Instincts, such as web spinning in spiders, might take hours or days to complete and are usually made up of many behaviors. The feeding behavior of an egg-eating snake is shown in **Figure 5.** The snake's pattern of behavior is an instinct.

 **Reading Check** Explain the difference between reflexes and instincts.

▲ **Figure 4** This armadillo has a reflex that causes it to jump when startled.

▭ **Review** **Personal Tutor**

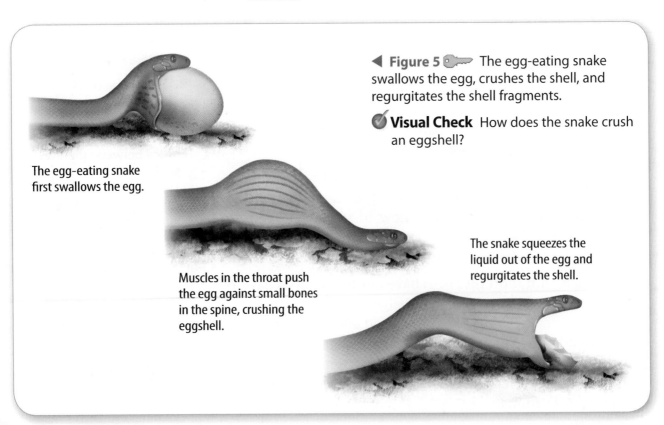

The egg-eating snake first swallows the egg.

Muscles in the throat push the egg against small bones in the spine, crushing the eggshell.

The snake squeezes the liquid out of the egg and regurgitates the shell.

◀ **Figure 5** ☞ The egg-eating snake swallows the egg, crushes the shell, and regurgitates the shell fragments.

**Visual Check** How does the snake crush an eggshell?

## Behavior Patterns

Many animal behaviors change in response to the change of seasons. In warm weather, there is plenty of food and water, and animals have no difficulty keeping warm. As the weather becomes cooler, food and water supplies might decrease, and animals might have difficulty surviving.

**Migration** Some animals move to warmer places during cooler weather. *This instinctive, seasonal movement of animals from one place to another is called* **migration.** Animals migrate to find food and water when the weather becomes too hot or too cold, or to return to specific breeding locations. Many birds, such as the ruby-throated hummingbird shown in **Figure 6**, fly many kilometers to warmer climates where they can find food.

**Figure 6** 🔑 Ruby-throated hummingbirds fly from New England to Louisiana. They then fly nonstop for about 805 km to the Yucatan Peninsula and Central and South America.

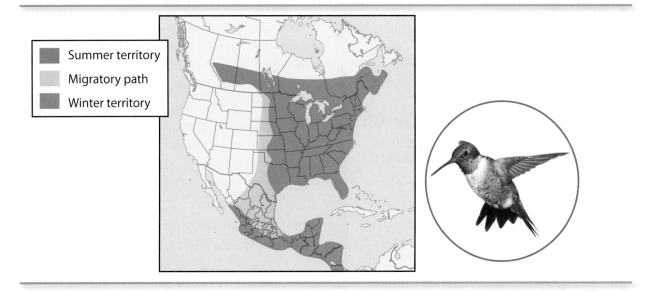

Summer territory
Migratory path
Winter territory

**Hibernation** Other animals do not migrate when temperatures get colder. Some animals, such as snowy owls and snowshoe hares, have feathers or fur that keep them warm in the winter. Other animals respond to cold temperatures and limited food supplies by hibernating. **Hibernation** *is a response in which an animal's body temperature, activity, heart rate, and breathing rate decrease during periods of cold weather.*

Chipmunks, some bat species, and prairie dogs are just a few types of animals that hibernate. Hibernating animals live on the fat that was stored in their bodies before hibernation. In some hibernating rodents, up to 50 percent of their body weight is fat.

Reptiles and other animals whose internal temperatures change with the environment do not hibernate. Rather, they enter a hibernationlike state. In dry, hot areas such as deserts, many animals also decrease their activity. This period of inactivity is called estivation (es tuh VAY shun).

**WORD ORIGIN** · · · · · · · · · · · ·

hibernation
from Latin *hibernare*, means "the action of passing the winter"

## Learned Behaviors

You have probably heard about service dogs that help humans by opening doors or turning on light switches. How are these dogs able to do such amazing things? Dogs and all other mammals, birds, reptiles, amphibians, and fish learn. This means that these animals develop new behaviors through experience or practice. Invertebrates, such as mollusks, insects, and arthropods, also can learn, but most of their behaviors are innate, or inherited.

### Imprinting

Young birds and mammals usually follow their mothers around. This helps protect them from danger and find food. How do they learn to do this? **Imprinting** *occurs when an animal forms an attachment to an organism or place within a specific time period after birth or hatching.* Once a young animal has imprinted itself on an organism, it will usually not attach itself to another. For example, a lamb might become imprinted on a human who fed it from a bottle. Once the lamb matures, it might have a hard time identifying as a member of a flock of sheep.

Not all imprinting occurs on organisms. Turtles do not imprint on other turtles. Female sea turtles return to the beach where they were born to lay their eggs. These turtles have imprinted on the beach.

### Trial and Error

Some behaviors, such as a child learning to button a shirt, take many tries before they are performed correctly. The child might try several buttoning techniques before finding one that works. This type of learning, called trial and error, happens in animals as well. For example, a monkey presented with food in a box might try to open the box many ways before succeeding. The next time it encounters a similar box, it will remember how to open the box without retrying the techniques that did not work.

---

**Inquiry MiniLab**　　　15 minutes

### How do young birds recognize predators?

When goslings, or baby geese, see a bird in the air that has a different wingspan or shape than the parent goose, they duck down. They seem to know they must not be seen by a predator.

1. Look at the pictures of the three birds in flight. In your Science Journal, describe the differences between each silhouette.

2. Choose at least two characteristics that are different for each bird.

**Analyze and Conclude**

1. **Evaluate** How could recognizing differences help a gosling survive?

2. 🔑 **Key Concept** Explain what type of behavior the gosling is exhibiting.

**452** ● Chapter 13

EXPLAIN

▲ Figure 7 🔑 Some fish learn through conditioning to come to the surface of the water when they are hungry.

🔑 **Visual Check**  What is the stimulus that the fish are responding to?

## Conditioning

Another way that animals might learn new behaviors is through conditioning. *In* **conditioning,** *behavior is modified so that a response to one stimulus becomes associated with a different stimulus.* As shown in **Figure 7,** some fish learn to come to the surface of the water when a hand is held over the water. They have learned that the hand often holds food. Through conditioning, some birds learn to avoid stinging wasps and monarch butterflies, which have a bad taste.

## Cognitive Behavior

Thinking, reasoning, and solving problems are cognitive behaviors. Humans use cognitive behavior to solve problems and plan for the future. Scientists have done experiments with animals such as primates, dolphins, elephants, and ravens that suggest they also might use cognitive behaviors. For example, studies done with ravens showed the birds could figure out how to get meat by pulling a string attached to the food. Other animals appear to show cognitive behaviors such as using tools to get food. For example, sea otters use rocks to crack the shells of clams and mussels, as shown in **Figure 8.**

🔑 **Key Concept Check**  How are animal behaviors classified?

Figure 8 🔑 Scientists have observed otters using what appears to be cognitive behavior. ▼

# Lesson 1 Review

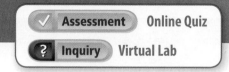
## Visual Summary

Animals react to stimuli with behaviors.

Behaviors can be either innate or learned.

Many animals have complex patterns of innate behaviors.

**FOLDABLES**

Use your lesson Foldable to review the lesson. Save your Foldable for the project at the end of the chapter.

## What do you think NOW?

You first read the statements below at the beginning of the chapter.

**1.** Animals react to their environments.

**2.** All animal behavior is instinctive.

Did you change your mind about whether you agree or disagree with the statements? Rewrite any false statements to make them true.

## Use Vocabulary

**1** **Define** *innate behavior* in your own words.

**2** **Use the term** *migration* in a sentence.

**3** **Distinguish** between conditioning and cognitive behavior.

## Understand Key Concepts

**4** Which is a learned behavior?
- **A.** conditioning
- **B.** instinct
- **C.** migration
- **D.** reflexes

**5** **Classify** the following behaviors as innate or learned: birds flying to warmer climates for winter, a mussel closing its shell, a duckling following its mother, a spider spinning a web.

**6** **Compare** learning by trial and error and conditioning.

## Interpret Graphics

**7** **Explain** Use the image below to explain how conditioning works in animals.

**8** **Identify** Copy and fill in the table below with examples of the types of behavior.

| Innate Behavior | Learned Behavior |
|---|---|
|  |  |
|  |  |
|  |  |
|  |  |

## Critical Thinking

**9** **Design an experiment** to determine if a goldfish can learn by conditioning.

# Can the color or surface of an area determine how a mealworm will move?

### Materials

mealworms

marker

metric ruler

string

**Also needed:**
surface materials

### Safety

Animal behaviors help maintain homeostasis, a steady internal state. An example of this is finding an environment that meets all the needs for living and survival, such as food, water, or shelter. If an animal is put in an environment that is not familiar and does not meet its needs, it will seek a more suitable environment.

## Learn It

In science experiments, the one factor you change is called the independent variable. The factor that changes as a result is the dependent variable. An experimental subject for which the independent variable is not changed is the control. By **manipulating variables,** you can get more accurate results from experiments.

## Try It

1. Read and complete a lab safety form.

2. Choose four different surfaces for testing movement of a mealworm. In your Science Journal, make a table like the one below to record your observations.

3. Carefully place worm 1 on the first surface, and leave it for 30 s. Trace the path of the worm as it moves across the surface. Use the string to follow the path it travels, then place the string on a ruler to measure the distance in centimeters.

4. Repeat step 3 two more times with the same worm.

5. Repeat steps 3 and 4 on different surfaces for worms 2, 3, and 4.

## Apply It

6. **Identify** the independent variable and the dependent variable. What factor enabled the mealworm to move the farthest? Explain.

7. **Summarize** Change another independent variable, such as putting one surface on an upward slant, putting flour, sand, or soil on one surface, or placing a transparent cover over the color of the surface. Summarize the outcomes from changing this independent variable.

8. 🔑 **Key Concept** What did the mealworms do to try to regain a balance with their environment after being moved from the container? What behavior is this?

| Worm | Trial | Surface | Distance (cm) |
|------|-------|---------|---------------|
| 1 | 1 | | |
| 1 | 2 | | |
| 1 | 3 | | |
| 2 | 1 | | |
| 2 | 2 | | |
| 2 | 3 | | |
| 3 | 1 | | |

# Lesson 2

## Reading Guide

### Key Concepts
**ESSENTIAL QUESTIONS**

- How do animals communicate?
- How do animals interact in societies?

### Vocabulary

**bioluminescence** p. 458

**pheromone** p. 459

**society** p. 460

**territory** p. 461

**aggression** p. 461

**g  Multilingual eGlossary**

# Interacting with Others

**Inquiry  Fighting or Playing?**

These red foxes appear to be fighting, but they are actually playing in the snow. All animals have ways to communicate and interact with other members of their species. They also have ways of communicating and interacting with other species.

# Launch Lab

**10 minutes**

## How are you feeling?

Although they do not use words, animals still can communicate. Many use sounds, such as chirping or singing, but some use only body language and facial expressions to communicate with other animals. Some animals even communicate changes in mood, all without words.

1. Brainstorm several emotions with your group, and write them in your Science Journal.

2. Decide which emotion you could demonstrate using just your facial expression and/or body language without props or touching anyone.

3. Take turns communicating your emotion and guessing which one each person is trying to show.

### Think About This

1. What was your emotion, and how did you communicate it?

2. How do you think animals might communicate that same emotion?

3. 🔑 **Key Concept** Why do you think animals might need to communicate with other animals?

## Communication

In the last lesson, you read about behaviors in individual animals. Animals have distinct behaviors in groups as well. Have you ever noticed a swarm of ants around a piece of food that has fallen on the sidewalk, as shown in **Figure 9?** How do you think the ants knew where to go? A foraging ant discovered the food and left a trail of chemicals for the other ants to follow. This and other types of communication are important for animal group behavior.

Animals use communication for many reasons, such as protection, locating other members of their groups, warning others of danger, and finding mates. Animals communicate using sound, light, chemicals, and body language. An animal might communicate with other animals of the same species, or it might communicate with different species in the same area.

🔑 **Key Concept Check** How do animals communicate?

**FOLDABLES**

Make a vertical two-tab book. Label it as shown. Use it to record what you learn about animal communication and animal societies.

Animal Communication

Animal Societies

**Figure 9** 🔑 Ants follow the trail of another ant from their colony to find food.

**Concepts in Motion** Animation

### How can you demonstrate sound communication?

Sometimes in the evening or during a walk in the woods, you hear a distinctive sound from an animal. It might be a bird with an unusual call, a cricket chirping, or a bullfrog croaking in a pond nearby. All of these sounds are forms of communication that enable animals to find each other.

1 Pull a **sound card** from the pile of cards, and wait until everyone is ready.

2 If your sound card directs you to make a sound, begin to make the sound as directed on the card. Continue to make the sound intermittently until your partner recognizes your call. If your sound card directs you to listen for a sound, listen carefully and find the student who is making the sound described on your card.

### Analyze and Conclude

1. **Explain** what made it possible for you to distinguish your partner from others who were also making sounds.

2. **Infer** What might happen if another person used a pattern or sound that was similar to the one you were listening for?

3. 🔑 **Key Concept** What other ways, besides sound, do animals in the wild use to find others of the same kind? Why do animals need to find others?

## Sound

Many animals, such as birds, amphibians, reptiles, and mammals, communicate with sound. Dolphins make a wide variety of sounds, including whistles and grunts. Each sound has a different meaning to the other dolphins, such as excitement, play, or warning of danger. Although many animals make calls, some animals, such as the ruffed grouse, produce sound in other ways. The male grouse makes a drumming noise to attract a mate by using its wings to beat the air. Many insects, such as cicadas and crickets, also produce sounds to attract mates.

**Figure 10** 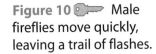 Male fireflies move quickly, leaving a trail of flashes.

## Light

To communicate in the dark, some animals use bioluminescence (BI oh lew muh NE sunts). **Bioluminescence** *is the ability of certain living things to give off light.* Chemical reactions in the animal's body produce the light. You might have seen fireflies, such as the one in **Figure 10,** as they blink out a code to attract females in the area. However, most animals that use bioluminescence live in the ocean. In the dimly lit zone of the ocean, up to 90 percent of fish and crustaceans are thought to use bioluminescence. Some fish use bioluminescence to lure prey into their mouths. Others have pockets of bioluminescent bacteria in their cheeks, which help the fish attract mates.

## Chemicals

Many animals produce chemicals, called pheromones (FER uh mohnz), to communicate. *A* **pheromone** *is a chemical that is produced by one animal and influences the behavior of another animal of the same species.* When released to the environment, pheromones can signal the presence of danger, food, mates, or even communicate the borders of a territory. Some flying animals, such as some moths, release pheromones into the air that attract mates. Other animals, such as male dogs, mark surfaces with pheromones that identify their territory to other dogs. Recall the ants that you read about in the beginning of this lesson. Ants leave a trail of one type of pheromone that leads other ants to food. They produce different pheromones that warn other ants of danger.

 **Reading Check** Summarize how animals use pheromones.

## Body Language

Can you tell what mood a person is in by looking at his or her face or body position? The person is using body language to communicate his or her mood. Animals also communicate with body language. Animals such as wolves communicate excitement, aggression, and other moods through facial expressions, as shown in **Figure 11.** Some parrots bob their heads when they are content and crouch with their heads down when they are sick or stressed. This body language can make it easier for an animal to communicate with other members of its species.

**WORD ORIGIN**

**pheromone**
from Greek *pherein*, means "to carry"

**Figure 11** Wolves communicate their moods through facial expressions and body language.

**Visual Check** If a wolf has narrowed eyes and ears laid back, what mood is it communicating?

## Wolf Body Language

**Aggression:**
• ears forward
• narrowed or staring eyes
• body tense and upright

**Playfulness:**
• ears relaxed
• wide open eyes
• relaxed body

**Fear:**
• ears laid back
• narrowed eyes
• body crouched low

Figure 12 🔑 In this hyena society, one adult member watches for danger while other group members feed.

# Societies and Behaviors

Have you ever seen a flock of birds flying together? Animals live in groups for many reasons, such as for protection and obtaining food. *A* **society** *is a group of animals of the same species living and working together in an organized way.* Animal societies are sometimes highly structured with specific roles for its members, such as that of the spotted hyenas shown in **Figure 12.** Spotted hyenas live in large groups of up to 90 members. The members work together to hunt and defend their kill. Other animal societies are less organized, and each member might serve different roles. Some species of animals group together closely only at certain times of the year, such as for breeding or migration.

## Dominance and Submission

Spotted hyena societies are organized by dominance. This means that the members are organized according to their social status relative to the other animals. The animal with the highest social status, the dominant animal, has power over the ones below it. Animals with a lower relative status to a dominant animal are **submissive** to that animal. In a spotted hyena society, females are most dominant, then cubs, and then males. Dominance also is important in groups of other animals, such as wolves, chickens, and some primates.

Dominance also might help reduce fighting among animals living in a society. For example, hyenas rarely hurt each other while fighting with other members of their society. Less dominant members usually submit to, or stop fighting, more dominant ones. Sometimes a submissive animal might mimic the behavior of a young animal and show that they are not a threat. For example, submissive wolves roll over or crouch, and less dominant hens move out of the way of the dominant hen.

## Territorial Behaviors

*Animals might set up and defend an area for feeding, mating, and raising young called a* **territory.** Some insects and most vertebrates have a territory. Animals might identify their territories by making noises, physically changing the territory by scraping bark off trees, or by marking the area with pheromones, urine, or feces.

Animals defend the borders of their territory from other members of their species. If the borders are crossed, the animal, such as the cat shown in **Figure 13,** first might attempt to scare or intimidate the invading animal. If the animal does not leave, the defender might use aggression. **Aggression** *is a forceful behavior used to dominate or control another animal.* When animals fight another member of the same species, they usually do not try to cause serious harm to the other animal. For example, giraffes have the ability to kick fiercely, and they use this ability to defend against predators such as lions. These attacks can be deadly. However, when two male giraffes show aggression towards each other, they push at each other with their necks. This behavior is common and rarely fatal.

## Courtship

Animals have specialized behaviors that help them find and attract a mate. They often compete with others of the same species for a mate. Some animals, such as female gypsy moths, release pheromones that attract males. Other animals, such as frogs and birds, use mating songs that gain the attention of mates. Some male birds bring the female a gift of food, such as a male tern bringing a fish to a female. Male fiddler crabs wave their enlarged claws and skitter across the ocean floor in the hopes of getting a female fiddler crab's attention. Male bowerbirds, such as the one shown in **Figure 14,** build elaborate nests using brightly colored objects during courtship.

 **Key Concept Check** How do animals interact in societies?

▲ **Figure 13** 🔑 When a cat puffs up its fur, it appears more threatening to intruders.

◀ **Figure 14** 🔑 Male bowerbirds use shiny or brightly colored objects while building nests in an attempt to attract a mate.

# Lesson 2 Review

## Visual Summary

Animals communicate in many ways, including with sound, light, chemicals, and body language.

Some animals live in societies that are highly structured.

Societal behaviors include dominance, submission, territorial behaviors, and courtship behaviors.

**FOLDABLES**

Use your lesson Foldable to review the lesson. Save your Foldable for the project at the end of the chapter.

## What do you think NOW?

You first read the statements below at the beginning of the chapter.

**3.** Some animals give off light to communicate with each other.

**4.** Animals always fight to protect their territories.

Did you change your mind about whether you agree or disagree with the statements? Rewrite any false statements to make them true.

## Use Vocabulary

**1** Some animals use _____ to communicate in the dark.

**2** **Define** *society* in your own words.

**3** **Use the term** *territory* in a complete sentence.

## Understand Key Concepts

**4** Which form of communication includes pheromones?
- **A.** chemicals
- **C.** body language
- **B.** light
- **D.** facial expressions

**5** **Compare** dominance and submission.

**6** **Infer** why light communication is common in marine environments.

## Interpret Graphics

**7** **Identify** the form of animal communication in the figure below. Explain what is happening.

**8** **Identify** Copy and fill in the graphic organizer below to identify the forms of animal communication.

Communication

## Critical Thinking

**9** **Hypothesize** what form of communication might be used by animals living in a noisy environment.

# Courtship
## Displays

### A "Superb" Way to Find a Mate

What kind of animal do you think this is? Although it might look like a cartoon character, these are pictures of a male superb bird of paradise during a courtship display.

A courtship display is a series of specialized behaviors that help an animal attract a mate. It can include movements, sounds, and/or chemical communication. In its courtship display, the male superb bird of paradise transforms itself, as shown in the following sequence.

**①** This pose is the beginning of the courtship display. At this stage, the bird sometimes moves its wings to make a clicking sound.

**②** In this part of the courtship display, the feathers on the top of the bird's head are displayed, and the blue feathers on the bird's chest are extended outward.

**③** The bird then extends the feathers on its head outward and around the head to form a rounded shape. Body feathers are extended as well. The bird appears as a black oval with a bright blue shape in the middle. The small blue spots are not the bird's eyes; they are spots on its feathers. In this stage of the display, the male bird hops around the female.

### It's Your Turn

**REPORT** Research another animal's courtship display. Draw diagrams to show the different appearances and behaviors that are part of the display. Share your research with your class.

▲ Male superb bird of paradise

# Animal Reproduction and Development

## Reading Guide

**Key Concepts**
ESSENTIAL QUESTIONS

- What are the roles of male and female reproductive organs?
- How do the two types of fertilization differ?
- What are the different types of animal development?

### Vocabulary

**sexual reproduction** p. 465

**testis** p. 466

**ovary** p. 466

**fertilization** p. 466

**zygote** p. 466

**metamorphosis** p. 470

 **Multilingual eGlossary**

 **Video**

- BrainPOP®
- What's Science Got to do With It?

## nquiry Leaving Home?

You probably know that caterpillars turn into butterflies. This butterfly is emerging from its chrysalis (KRIH suh lus). Why do you think the different life stages of this animal look so different? What other types of animal development are there?

## Launch Lab

**15 minutes**

### How is development similar in different animals?

No matter what the life span is of any animal, it starts with birth and develops to maturity, which is called adulthood. No matter the size of the animal, you might be able to see some similarities in the young animals.

1 Examine the pictures of young animals and adults in the table.

2 In your Science Journal, note some similarities between each baby and the adult counterpart.

3 Find one thing the young have in common and one thing the adults have in common.

**Think About This**

1. What similarities were you able to find between young animals and adults?

2. What similar characteristics were you able to identify in the babies as a group?

3. 🔑 **Key Concept** What do you think was the most obvious characteristic related to development in all the animals pictured? Explain.

## Sexual Reproduction

Have you ever found a cluster of tiny, beadlike structures on the underside of a leaf? They might be eggs laid by a butterfly, a ladybug, or some other insect. Eggs are an important part of the life cycle of many animals. *In* **sexual reproduction**, *the genetic material from two different cells—a sperm and an egg—combine, producing an offspring.* Most animals reproduce sexually, although some can reproduce asexually, without a sperm and egg joining.

Male and female animals of the same species often look different from each other. It's easy to tell the difference between the male and the female shown in **Figure 15.** In mammals and birds, males are often larger or more colorful than females.

**ACADEMIC VOCABULARY**

cycle
*(noun)* a series of events that regularly recur and lead back to the starting point

**Figure 15** A lion has a ruff of fur around his neck and is larger than the lioness.

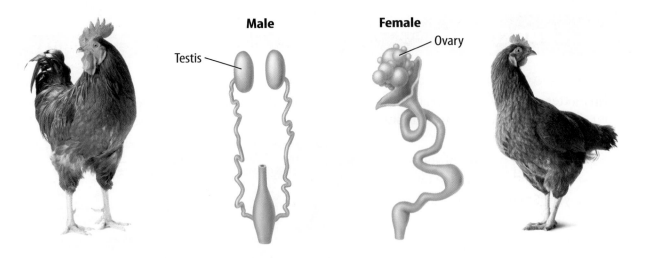

Male
Testis

Female
Ovary

**Figure 16**  Testes of male animals produce sperm cells. Ovaries of female animals produce egg cells.

## Use Ratios

A ratio can be used to compare data about sperm and eggs. For example, the head of a human sperm cell averages 5 μm. (A μm, or micron, is one-millionth of a meter.) If the tail of the sperm cell measures 50 μm, what is the ratio of tail to head?

Set up the two numbers as a ratio by writing them in any of the following forms:

50 to 5; 50:5; $\frac{50}{5}$

Reduce the numbers to their lowest form.

Divide each side by 5. The ratio is 10 to 1 or 10:1 or $\frac{10}{1}$

## Practice
A fruit fly sperm cell is about 1.8 mm in length. What is the ratio of a human sperm cell to a fly sperm cell?

  **Review**

- **Math Practice**
- **Personal Tutor**

## Male Reproductive Organs

The reproductive system of an animal includes specialized reproductive organs that produce sperm or eggs. The reproductive systems of male and female chickens are shown in **Figure 16.** Male animals have **testes** (TES teez; singular, testis), *the male reproductive organs that produce sperm.* Sperm are reproductive cells with tails that enable them to swim through fluid to reach an egg cell. Most male animals have two testes located inside the body cavity.

## Female Reproductive Organs

Female animals have **ovaries,** (OH va reez) *the female reproductive organs that produce egg cells.* Most female animals have two ovaries except female birds, such as the chicken in **Figure 16,** which have only one ovary. Egg cells are larger than sperm and cannot move on their own.

**Key Concept Check** What are the functions of male and female reproductive organs?

## Fertilization

Sexual reproduction requires **fertilization**—*the joining of an egg cell with a sperm cell.* Half of the genetic material is in a sperm cell. The other half is in an egg cell. *When a sperm cell fertilizes an egg cell, the new cell that forms is called a* **zygote** (ZI goht). The zygote develops into a new organism and contains the genetic material from both the sperm cell and the egg cell.

Not all animals fertilize their eggs in the same way. The eggs of some animals are fertilized inside the mother's body, and some are fertilized outside the body. Next, you will read about both types of fertilization and how organisms develop.

**Internal Fertilization** When fertilization occurs inside the body of an animal, it is called internal fertilization. For many animals, the male has a specialized structure that can deposit sperm in or near the female's reproductive system. The sperm swim to the egg or eggs. Earthworms, spiders, insects, reptiles, birds, and mammals have internal fertilization.

Internal fertilization ensures that an embryo, which develops from a fertilized egg, or zygote, is protected and nourished until it leaves the female's body. This increases the chance that an embryo will survive, develop into an adult, and reproduce.

**External Fertilization** A female frog, such as the one shown in **Figure 17,** deposits unfertilized eggs under water. A male frog releases its sperm above the eggs as the female lays them. Fertilization that occurs outside the body of an animal is called external fertilization. When a sperm reaches an egg, fertilization takes place. Animals that reproduce using external fertilization include jellyfish, clams, sea urchins, sea stars, many species of fish, and amphibians.

Most animals that reproduce using external fertilization do not care for the fertilized eggs or for the newly hatched young. As a result, eggs and young are exposed to predators and other dangers in the environment, reducing their chances of surviving. Successful reproduction of animals with external fertilization requires that a large number of eggs be produced to ensure that at least a few offspring will survive to become adults.

 **Key Concept Check** How do the two types of fertilization differ?

**FOLDABLES**

Make a vertical three-tab book. Draw a Venn diagram on the front before cutting the folds to form three tabs. Label it as shown. Use it to compare internal and external fertilization.

Internal Fertilization

Both

External Fertilization

**Figure 17** For organisms that have external fertilization, mating behaviors help make certain that eggs are fertilized as soon as possible after they leave the female's body.

## Development

The zygote produced by fertilization is only the beginning of an animal's development. The zygote grows by mitosis and cell divisions and becomes an embryo—the next stage in an animal's development. A growing embryo needs nourishment and protection from predators and other dangers in the environment. Different animals have different ways of supplying the needs of an embryo. In some animals, the embryo develops outside the mother's body. In others, the embryo develops inside the mother.

### External Development

The grass snake in **Figure 18** is an example of an animal whose embryos develop outside the mother. Animals that develop outside the mother usually are protected inside an egg. In most instances, one embryo develops inside each egg. Most eggs contain a yolk that provides food for the developing embryo. A covering surrounds the egg. The covering protects the embryo, helps keep it moist, and discourages predators. Eggs laid by lizards, snakes, and other reptiles have a tough, leathery covering, as shown in **Figure 18.** A tough jellylike substance usually surrounds eggs laid under water, such as those laid by frogs. Bird eggs have a hard covering called a shell.

**Figure 18** 🔑 This grass snake protects its eggs while the embryos develop.

## Internal Development

The embryos of some animals, including most mammals, develop inside the mother. These embryos get nourishment from the mother. An organ or tissue transfers nourishment from the mother to the embryo. Other embryos, such as those of some snakes, insects, and fish, develop in an egg with a yolk while inside the mother. For these animals, the yolk, not the mother, provides nourishment for the developing young. The young hatch from the eggs while they are inside the mother and then leave the mother's body.

 **Reading Check** Where does a snake embryo get nourishment if it develops inside its mother?

## Gestation

The length of time between fertilization and birth of an animal is called the gestation (jeh STAY shun) period. Gestation period varies from species to species and usually relates to the size of an animal—the smaller the animal, the shorter its gestation period. For example, the gestation period for a mouse is about 21 days, for a human is about 266 days, and for an elephant is about 600 days. The gestation period for a kangaroo is 35 days. A kangaroo is only about 2.5 cm long at birth, as shown in **Figure 19**. Most of its development occurs in a pouch on the mother's body.

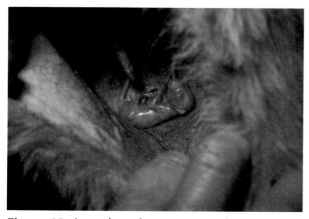

**Figure 19** A newborn kangaroo crawls into a pouch on the mother's body. It feeds and grows inside the pouch until it is large enough to live on its own.

## Inquiry MiniLab
### 15 minutes

### Is it possible to select which animal will have the largest newborn if you know the gestation period?

If you looked at pictures of newborn animals and compared them to a set of pictures of the animal mothers, you would probably say that the larger baby had a larger mother. What do you think you would find if you compared the gestation period of these animals?

1. Analyze the data in the table below.

2. As you are making comparisons using the data, see if you recognize any numbers that seem unusual. Record these in your Science Journal.

3. Graph your data on a line graph.

| Animal | Gestation Period in Days (average) | Newborn Weight in kg (average) |
|---|---|---|
| Meadow mouse | 18 | 0.0008 |
| Guinea pig | 68 | 0.1 |
| Porcupine | 105 | 0.2 |
| Giant panda | 135 | 0.2 |
| Squirrel monkey | 150 | 0.1 |
| Ribbon seal | 330 | 10.5 |
| Bactrian camel | 406 | 40 |
| Giraffe | 435 | 70 |
| Elephant | 660 | 113 |

### Analyze and Conclude

1. **Identify** the animal(s) that had a gestation period or newborn weight that did not follow a logical pattern.

2. **Explain** how the discrepancy was shown on the graph.

3. 🔑 **Key Concept** Can you make a comparison between the size of an animal, the length of the gestation period, and the weight of the newborn?

**WORD ORIGIN** ········

**metamorphosis**
from Greek *meta*, means
"change"; and *morphe*, means
"form"

## Metamorphosis

Some animals, including amphibians and many animals without backbones, go through more than one phase of development. **Metamorphosis** (me tuh MOR fuh sihs) *is a developmental process in which the form of the body changes as an animal grows from an egg to an adult.*

The metamorphosis of a ladybug and the metamorphosis of a frog are shown in **Figure 20.** During its development, the ladybug goes from egg to larva to pupa to adult. The tadpole is the larval stage of a frog. Larva and adult forms often have different lifestyles. The larva of the frog lives only in water. The adult frog can live on land or in water.

 **Key Concept Check** What are the different types of animal development?

**Ladybug Life Cycle**

Concepts in Motion Animation

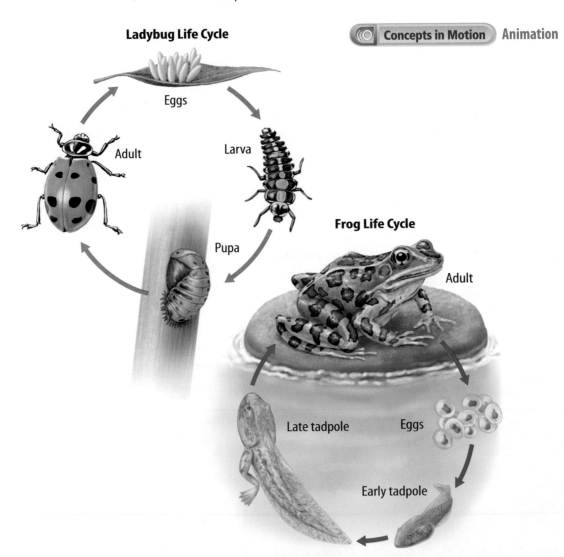

**Frog Life Cycle**

**Figure 20** A ladybug larva hatches from the egg, changes to a pupa, then the pupa changes into an adult. A tadpole hatches from an egg. It grows legs and loses its tail as it develops into an adult frog.

**Visual Check** Which developmental stage is not in a frog life cycle?

# Lesson 3 Review

## Visual Summary

Most animals reproduce sexually, and male and female animals often look different.

Fertilization can be internal or external.

Some animals have internal development, and others have external development.

FOLDABLES

Use your lesson Foldable to review the lesson. Save your Foldable for the project at the end of the chapter.

## What do you think NOW?

You first read the statements below at the beginning of the chapter.

**5.** During sexual reproduction, a sperm cell and an egg cell join.

**6.** Some animals develop inside the mother.

Did you change your mind about whether you agree or disagree with the statements? Rewrite any false statements to make them true.

## Use Vocabulary

**1** **Distinguish** between testes and ovaries.

**2** **Define** *metamorphosis* in your own words.

**3** The production of offspring by joining of a sperm and an egg is called _____.

## Understand Key Concepts

**4** Which are the reproductive cells that form in female animals?
   A. eggs        C. sperm
   B. ovaries     D. testes

**5** **Infer** why snake eggs have leathery shells.

**6** **Contrast** the survival of offspring from species with internal or external fertilization.

## Interpret Graphics

**7** **Explain** Use the image below to explain the benefits of external fertilization in frogs.

**8** **Sequence** Copy and fill in the graphic organizer below to sequence the stages of metamorphosis in a ladybug.

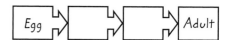

## Critical Thinking

**9** **Hypothesize** why large animals have a longer period of gestation than small animals.

## Math Skills ✕÷＋        Review
—————————— Math Practice ——

**10** A human egg cell has a diameter of about 120 μm. If a human sperm cell measures 5 μm, what is the ratio of the size of the egg cell to the size of the sperm cell?

## Materials

rectangular plastic container with lid

construction paper

earthworms

gooseneck lamp

paper towels

sand

dark soil

## Safety

# What changes an earthworm's behavior?

Have you ever seen an earthworm on the ground after it rains? Earthworms favor moist conditions and often are found in gardens or forest soil. They move to more favorable conditions when their environment becomes unsuitable.

## Ask a Question

Think about ways you could investigate an earthworm without hurting the worm. Develop a question based on your thoughts. If you want to be sure your question is testable, consider the variables, constants, and equipment that would be involved.

## Make Observations

1 Read and complete a lab safety form.

2 Observe the earthworms in your container, and think about their needs.

3 In your Science Journal, write down some ideas you could easily explore about earthworm behavior.

4 Discuss your ideas with your group, and choose one idea. Identify your variables and your control.

5 Ask your teacher for approval of your plan and any materials that you might need that are not available already.

6 Set up the lab materials according to your plan.

## Form a Hypothesis

**7** After you have looked over your plan and lab setup, discuss what you think you will find out about the earthworm in response to the stimulus you chose. Form a hypothesis to explain the relationship between the change in the environment and the earthworm's behavior.

## Test Your Hypothesis

**8** Make adjustments, if necessary, to your lab setup and get one or more worms from your container.

**9** Decide how you are going to record your observations, and create a table in your Science Journal.

**10** Follow your plan and record your observations.

## Analyze and Conclude

**11** **Compare** the behavior of the earthworm before and after you applied the stimulus.

**12** **Interpret** any unexpected responses during one or more trials.

**13** **Infer** from your data if the earthworm learned to change its behavior because of repeated trials.

**14** **The Big Idea** What do you think would happen if a worm in a natural environment encountered the change you designed?

## Communicate Your Results

Draw a comic strip depicting your question, your hypothesis, and the results. Share your comic strip with the class.

 **Extension**

Observe both mealworms and earthworms under the same conditions. Predict which worms would get used to the new environment faster.

Ⓜ **10**

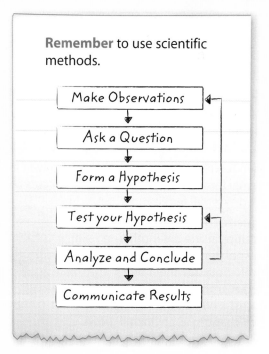

**Lab Tips**

☑ Be gentle when handling the worms.

☑ Take care to keep hot objects such as lightbulbs away from the animals.

**Remember** to use scientific methods.

> Make Observations
> Ask a Question
> Form a Hypothesis
> Test your Hypothesis
> Analyze and Conclude
> Communicate Results

# Chapter 13 Study Guide

**THE BIG IDEA**

Animals communicate using sound, light, chemicals, and body language. Societies and social behaviors help them interact with each other. Courtship behaviors help animals find mates. Most animals use sexual reproduction.

| Key Concepts Summary 🔑 | Vocabulary |
|---|---|
| **Lesson 1: Types of Behavior**<br>• Animal **behaviors** help maintain homeostasis by reacting to stimuli in their internal and external environments.<br>• Animal behaviors can be **innate** or learned.<br><br> | **behavior** p. 447<br>**innate behavior** p. 449<br>**instinct** p. 450<br>**migration** p. 451<br>**hibernation** p. 451<br>**imprinting** p. 452<br>**conditioning** p. 453 |
| **Lesson 2: Interacting with Others**<br><br>• Animals use sound, light, chemicals, and body language to communicate.<br>• Animals live and work together in **societies.** They might exhibit dominance, submission, territorial behaviors, and courtship. | **bioluminescence** p. 458<br>**pheromone** p. 459<br>**society** p. 460<br>**territory** p. 461<br>**aggression** p. 461 |
| **Lesson 3: Animal Reproduction and Development**<br>• Male reproductive organs, called **testes,** produce sperm. Female reproductive organs, called **ovaries,** produce eggs. In **sexual reproduction,** the egg and the sperm join to form a new organism.<br>• When **fertilization** occurs inside the body of an animal, it is called internal fertilization. Fertilization that occurs outside the body is called external fertilization.<br>• In internal development, the embryo develops inside the mother. In external development, embryos develop outside the mother. **Metamorphosis** is a developmental process in which the form of the body changes as an animal grows from an egg to an adult.<br> | **sexual reproduction** p. 465<br>**testis** p. 466<br>**ovary** p. 466<br>**fertilization** p. 466<br>**zygote** p. 466<br>**metamorphosis** p. 470 |

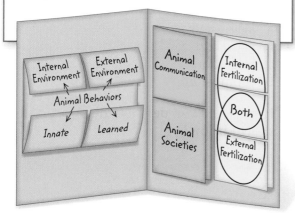
## FOLDABLES® Chapter Project

Assemble your lesson Foldables as shown to make a Chapter Project. Use the project to review what you have learned in this chapter.

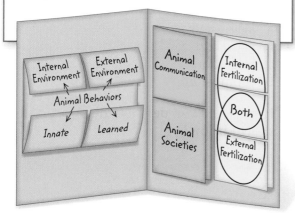

Internal Environment    External Environment

Animal Behaviors

Innate    Learned

Animal Communication

Animal Societies

Internal Fertilization

Both

External Fertilization

## Use Vocabulary

1. The way an organism reacts to other organisms is _____.

2. Define the term *hibernation* in your own words.

3. In chemical communication, an animal might use a(n) _____.

4. Use the word *aggression* in a complete sentence.

5. The joining of a sperm cell and an egg cell is called _____.

6. Use the word *zygote* in a sentence.

## Link Vocabulary and Key Concepts

**Concepts in Motion**   Interactive Concept Map

*Copy this concept map, and then use vocabulary terms from the previous page and other terms from the chapter to complete the concept map.*

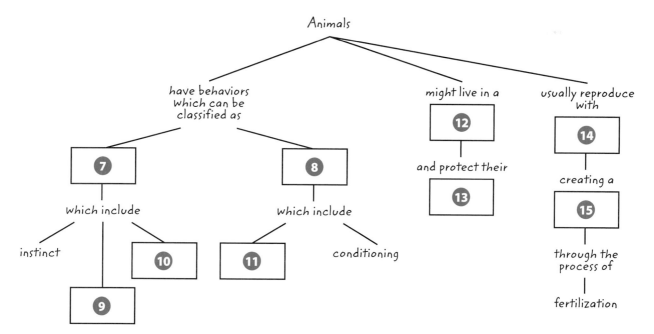

Animals

have behaviors which can be classified as

might live in a **12**

usually reproduce with **14**

**7**

**8**

and protect their **13**

creating a **15**

which include

which include

through the process of

instinct

**10**

**11**

conditioning

fertilization

**9**

## Understand Key Concepts

1. Which is a reflex?
   A. a bird building a nest
   B. pulling a string to get food
   C. pupils getting smaller in dim light
   D. tying your shoelaces

2. Which animal does NOT hibernate?
   A. bat
   B. chipmunk
   C. snake
   D. squirrel

3. Which type of animal behavior is shown in the figure below?

   A. conditioning
   B. imprinting
   C. instinct
   D. reflex

4. Which is a learned response that uses reasoning from past experiences?
   A. conditioning
   B. imprinting
   C. cognitive behavior
   D. trial and error

5. Body language is an example of
   A. communication.
   B. conditioning.
   C. migration.
   D. societies.

6. Which type of behavior is shown in the figure below?

   A. aggression
   B. courtship
   C. migration
   D. submission

7. How would you describe the organ system shown below?

   A. asexual
   B. embryo
   C. female
   D. male

8. What is the length of time between fertilization and birth called?
   A. external development
   B. gestation period
   C. metamorphosis
   D. zygote

9. What are the reproductive cells that form in male animals?
   A. eggs
   B. ovaries
   C. sperm
   D. testes

## Critical Thinking

**10 Formulate** a question to ask a scientist who investigated cognitive behavior in ravens.

**11 Summarize** how animals respond to change.

**12 Develop** a plan of communication for an animal that lives in darkness.

**13 Hypothesize** how the bird shown below might attract more mates.

**14 Research** one example of body language for a tree-dwelling mammal.

**15 Hypothesize** why most frog species enter water to reproduce.

**16 Compare** the number of eggs produced by animals that reproduce by external fertilization with the number of eggs produced by animals that reproduce by internal fertilization.

**17 Consider** why mammals do not develop by metamorphosis.

*Writing in Science*

**18 Write** a five-sentence paragraph comparing internal and external fertilization. Be sure to include a topic sentence and a concluding sentence in your paragraph.

## REVIEW THE BIG IDEA

**19** How do animals communicate, interact, and reproduce?

**20** What type of behavior is illustrated in the photo below?

## Math Skills

Review — Math Practice

### Use Ratios

**21** In one species of trout, the egg cell has a diameter of 5 mm. The circumference of the cell is 15.7 mm. What is the ratio of the circumference to the diameter of the cell?

**22** A chinook salmon has an average body length of 70 cm. If the sperm cell measures 55 μm, what is the ratio of the length of the salmon's body to the length of the sperm cell? [Hint: 1 cm = 10,000 μm]

**23** A sperm cell contains genetic material with a mass of 1.26 picogram (pg) [1 pg = one-trillionth of a gram]. After the sperm fertilizes an egg, the zygote contains 2.52 pg of genetic material. What is the ratio of genetic material from the egg and the sperm cells? [Hint: Subtract the genetic material of the sperm from the total to get the egg's contribution.]

*Record your answers on the answer sheet provided by your teacher or on a sheet of paper.*

## Multiple Choice

**1** Which is NOT one of the ways that animals communicate?

   **A** They make chemicals.

   **B** They migrate long distances.

   **C** They use light.

   **D** They use sound.

**2** A wolf rolls onto its back, exposing its belly to its pack mates. Which behavior does this show?

   **A** aggression

   **B** courtship

   **C** dominance

   **D** submission

*Use the figure below to answer question 3.*

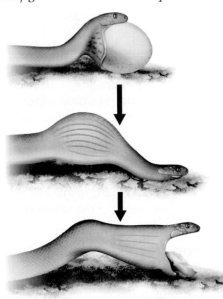

**3** Which term describes the behavior shown in the figure?

   **A** conditioning

   **B** imprinting

   **C** instinct

   **D** reflex

*Use the figure below to answer question 4.*

**4** Which are produced in the structure shown in the figure?

   **A** embryos

   **B** zygotes

   **C** egg cells

   **D** sperm cells

**5** Which is true of external fertilization?

   **A** It happens outside the female body.

   **B** It involves a small number of eggs.

   **C** It occurs only in the spring months.

   **D** It requires extended parental care.

**6** When a baby kangaroo is born, it crawls into its mother's pouch. Which type of behavior is this?

   **A** cognitive behavior

   **B** imprinting

   **C** innate behavior

   **D** trial and error

**7** A turtle perches on a log in the sun. How does this behavior help the turtle maintain homeostasis?

   **A** It attracts suitable mates.

   **B** It frightens potential predators.

   **C** It maintains body temperature.

   **D** It protects newborn offspring.

*Use the figure below to answer questions 8 and 9.*

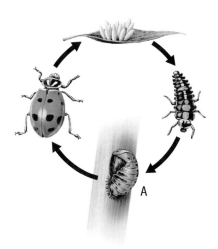

**8** Which stage of development is marked *A* in the figure?

   **A** adult

   **B** larva

   **C** pupa

   **D** zygote

**9** Which process occurs at the stage marked *A* in the figure?

   **A** fertilization

   **B** imprinting

   **C** hibernation

   **D** metamorphosis

**10** Which is the result of fertilization?

   **A** egg

   **B** embryo

   **C** sperm

   **D** zygote

## Constructed Response

*Use the figure below to answer questions 11 and 12.*

**11** Describe the behavior shown in the figure. Give an example of a stimulus that could cause the behavior shown.

**12** Explain how the behavior shown in the figure is an example of the flight-or-fight response.

**13** Describe how animals use chemicals as a territorial behavior.

**14** A shopping mall was constructed where a forest once grew. Many of the animals living in the area moved to another wooded area. Is this movement of animals an example of migration? Why or why not?

| NEED EXTRA HELP? | | | | | | | | | | | | | | |
|---|---|---|---|---|---|---|---|---|---|---|---|---|---|---|
| If You Missed Question... | 1 | 2 | 3 | 4 | 5 | 6 | 7 | 8 | 9 | 10 | 11 | 12 | 13 | 14 |
| Go to Lesson... | 2 | 2 | 1 | 1 | 3 | 1 | 1 | 3 | 3 | 3 | 1 | 1 | 2 | 1 |

# Student Resources

## For Students and Parents/Guardians

These resources are designed to help you achieve success in science. You will find useful information on laboratory safety, math skills, and science skills. In addition, science reference materials are found in the Reference Handbook. You'll find the information you need to learn and sharpen your skills in these resources.

# Table of Contents

# Scientific Methods

Scientists use an orderly approach called the scientific method to solve problems. This includes organizing and recording data so others can understand them. Scientists use many variations in this method when they solve problems.

## Identify a Question

The first step in a scientific investigation or experiment is to identify a question to be answered or a problem to be solved. For example, you might ask which gasoline is the most efficient.

## Gather and Organize Information

After you have identified your question, begin gathering and organizing information. There are many ways to gather information, such as researching in a library, interviewing those knowledgeable about the subject, and testing and working in the laboratory and field. Fieldwork is investigations and observations done outside of a laboratory.

**Researching Information** Before moving in a new direction, it is important to gather the information that already is known about the subject. Start by asking yourself questions to determine exactly what you need to know. Then you will look for the information in various reference sources, like the student is doing in **Figure 1.** Some sources may include textbooks, encyclopedias, government documents, professional journals, science magazines, and the Internet. Always list the sources of your information.

**Figure 1** The Internet can be a valuable research tool.

**Evaluate Sources of Information** Not all sources of information are reliable. You should evaluate all of your sources of information, and use only those you know to be dependable. For example, if you are researching ways to make homes more energy efficient, a site written by the U.S. Department of Energy would be more reliable than a site written by a company that is trying to sell a new type of weatherproofing material. Also, remember that research always is changing. Consult the most current resources available to you. For example, a 1985 resource about saving energy would not reflect the most recent findings.

Sometimes scientists use data that they did not collect themselves, or conclusions drawn by other researchers. This data must be evaluated carefully. Ask questions about how the data were obtained, if the investigation was carried out properly, and if it has been duplicated exactly with the same results. Would you reach the same conclusion from the data? Only when you have confidence in the data can you believe it is true and feel comfortable using it.

SCIENCE SKILL HANDBOOK

MATH SKILL HANDBOOK

FOLDABLES HANDBOOK

REFERENCE HANDBOOK

GLOSSARY/ GLOSARIO

INDEX

**Interpret Scientific Illustrations** As you research a topic in science, you will see drawings, diagrams, and photographs to help you understand what you read. Some illustrations are included to help you understand an idea that you can't see easily by yourself, like the tiny particles in an atom in **Figure 2.** A drawing helps many people to remember details more easily and provides examples that clarify difficult concepts or give additional information about the topic you are studying. Most illustrations have labels or a caption to identify or to provide more information.

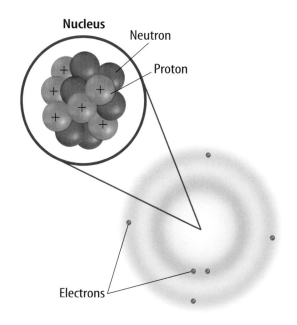

**Figure 2** This drawing shows an atom of carbon with its six protons, six neutrons, and six electrons.

**Concept Maps** One way to organize data is to draw a diagram that shows relationships among ideas (or concepts). A concept map can help make the meanings of ideas and terms more clear, and help you understand and remember what you are studying. Concept maps are useful for breaking large concepts down into smaller parts, making learning easier.

**Network Tree** A type of concept map that not only shows a relationship, but how the concepts are related is a network tree, shown in **Figure 3.** In a network tree, the words are written in the ovals, while the description of the type of relationship is written across the connecting lines.

When constructing a network tree, write down the topic and all major topics on separate pieces of paper or notecards. Then arrange them in order from general to specific. Branch the related concepts from the major concept and describe the relationship on the connecting line. Continue to more specific concepts until finished.

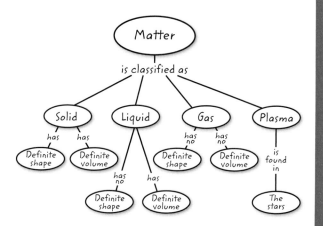

**Figure 3** A network tree shows how concepts or objects are related.

**Events Chain** Another type of concept map is an events chain. Sometimes called a flow chart, it models the order or sequence of items. An events chain can be used to describe a sequence of events, the steps in a procedure, or the stages of a process.

When making an events chain, first find the one event that starts the chain. This event is called the initiating event. Then, find the next event and continue until the outcome is reached, as shown in **Figure 4** on the next page.

SCIENCE SKILL HANDBOOK

MATH SKILL HANDBOOK

FOLDABLES HANDBOOK

REFERENCE HANDBOOK

GLOSSARY/ GLOSARIO

INDEX

SCIENCE SKILL HANDBOOK

MATH SKILL HANDBOOK

FOLDABLES HANDBOOK

REFERENCE HANDBOOK

GLOSSARY/ GLOSARIO

INDEX

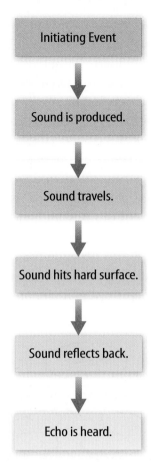

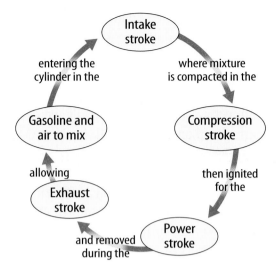

**Figure 4** Events-chain concept maps show the order of steps in a process or event. This concept map shows how a sound makes an echo.

**Figure 5** A cycle map shows events that occur in a cycle.

**Cycle Map** A specific type of events chain is a cycle map. It is used when the series of events do not produce a final outcome, but instead relate back to the beginning event, such as in **Figure 5.** Therefore, the cycle repeats itself.

To make a cycle map, first decide what event is the beginning event. This is also called the initiating event. Then list the next events in the order that they occur, with the last event relating back to the initiating event. Words can be written between the events that describe what happens from one event to the next. The number of events in a cycle map can vary, but usually contain three or more events.

**Spider Map** A type of concept map that you can use for brainstorming is the spider map. When you have a central idea, you might find that you have a jumble of ideas that relate to it but are not necessarily clearly related to each other. The spider map on sound in **Figure 6** shows that if you write these ideas outside the main concept, then you can begin to separate and group unrelated terms so they become more useful.

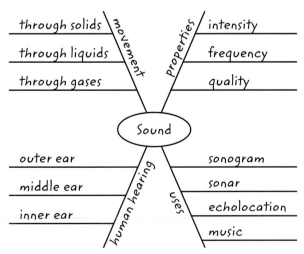

**Figure 6** A spider map allows you to list ideas that relate to a central topic but not necessarily to one another.

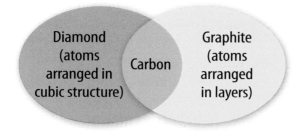

**Figure 7** This Venn diagram compares and contrasts two substances made from carbon.

**Venn Diagram** To illustrate how two subjects compare and contrast you can use a Venn diagram. You can see the characteristics that the subjects have in common and those that they do not, shown in **Figure 7.**

To create a Venn diagram, draw two overlapping ovals that are big enough to write in. List the characteristics unique to one subject in one oval, and the characteristics of the other subject in the other oval. The characteristics in common are listed in the overlapping section.

**Make and Use Tables** One way to organize information so it is easier to understand is to use a table. Tables can contain numbers, words, or both.

To make a table, list the items to be compared in the first column and the characteristics to be compared in the first row. The title should clearly indicate the content of the table, and the column or row heads should be clear. Notice that in **Table 1** the units are included.

| Table 1 Recyclables Collected During Week | | | |
|---|---|---|---|
| **Day of Week** | **Paper (kg)** | **Aluminum (kg)** | **Glass (kg)** |
| **Monday** | 5.0 | 4.0 | 12.0 |
| **Wednesday** | 4.0 | 1.0 | 10.0 |
| **Friday** | 2.5 | 2.0 | 10.0 |

**Make a Model** One way to help you better understand the parts of a structure, the way a process works, or to show things too large or small for viewing is to make a model. For example, an atomic model made of a plastic-ball nucleus and chenille stem electron shells can help you visualize how the parts of an atom relate to each other. Other types of models can be devised on a computer or represented by equations.

## Form a Hypothesis

A possible explanation based on previous knowledge and observations is called a hypothesis. After researching gasoline types and recalling previous experiences in your family's car you form a hypothesis—our car runs more efficiently because we use premium gasoline. To be valid, a hypothesis has to be something you can test by using an investigation.

**Predict** When you apply a hypothesis to a specific situation, you predict something about that situation. A prediction makes a statement in advance, based on prior observation, experience, or scientific reasoning. People use predictions to make everyday decisions. Scientists test predictions by performing investigations. Based on previous observations and experiences, you might form a prediction that cars are more efficient with premium gasoline. The prediction can be tested in an investigation.

**Design an Experiment** A scientist needs to make many decisions before beginning an investigation. Some of these include: how to carry out the investigation, what steps to follow, how to record the data, and how the investigation will answer the question. It also is important to address any safety concerns.

SCIENCE SKILL HANDBOOK

MATH SKILL HANDBOOK

FOLDABLES HANDBOOK

REFERENCE HANDBOOK

GLOSSARY/ GLOSARIO

INDEX

SCIENCE SKILL HANDBOOK

MATH SKILL HANDBOOK

FOLDABLES HANDBOOK

REFERENCE HANDBOOK

GLOSSARY/ GLOSARIO

INDEX

## Test the Hypothesis

Now that you have formed your hypothesis, you need to test it. Using an investigation, you will make observations and collect data, or information. This data might either support or not support your hypothesis. Scientists collect and organize data as numbers and descriptions.

**Follow a Procedure** In order to know what materials to use, as well as how and in what order to use them, you must follow a procedure. **Figure 8** shows a procedure you might follow to test your hypothesis.

| Procedure | |
|---|---|
| **Step 1** | Use regular gasoline for two weeks. |
| **Step 2** | Record the number of kilometers between fill-ups and the amount of gasoline used. |
| **Step 3** | Switch to premium gasoline for two weeks. |
| **Step 4** | Record the number of kilometers between fill-ups and the amount of gasoline used. |

**Figure 8** A procedure tells you what to do step-by-step.

**Identify and Manipulate Variables and Controls** In any experiment, it is important to keep everything the same except for the item you are testing. The one factor you change is called the independent variable. The change that results is the dependent variable. Make sure you have only one independent variable, to assure yourself of the cause of the changes you observe in the dependent variable. For example, in your gasoline experiment the type of fuel is the independent variable. The dependent variable is the efficiency.

Many experiments also have a control—an individual instance or experimental subject for which the independent variable is not changed. You can then compare the test results to the control results. To design a control you can have two cars of the same type. The control car uses regular gasoline for four weeks. After you are done with the test, you can compare the experimental results to the control results.

## Collect Data

Whether you are carrying out an investigation or a short observational experiment, you will collect data, as shown in **Figure 9.** Scientists collect data as numbers and descriptions and organize them in specific ways.

**Observe** Scientists observe items and events, then record what they see. When they use only words to describe an observation, it is called qualitative data. Scientists' observations also can describe how much there is of something. These observations use numbers, as well as words, in the description and are called quantitative data. For example, if a sample of the element gold is described as being "shiny and very dense" the data are qualitative. Quantitative data on this sample of gold might include "a mass of 30 g and a density of 19.3 $g/cm^3$."

**Figure 9** Collecting data is one way to gather information directly.

**Figure 10** Record data neatly and clearly so it is easy to understand.

When you make observations you should examine the entire object or situation first, and then look carefully for details. It is important to record observations accurately and completely. Always record your notes immediately as you make them, so you do not miss details or make a mistake when recording results from memory. Never put unidentified observations on scraps of paper. Instead they should be recorded in a notebook, like the one in **Figure 10.** Write your data neatly so you can easily read it later. At each point in the experiment, record your observations and label them. That way, you will not have to determine what the figures mean when you look at your notes later. Set up any tables that you will need to use ahead of time, so you can record any observations right away. Remember to avoid bias when collecting data by not including personal thoughts when you record observations. Record only what you observe.

**Estimate** Scientific work also involves estimating. To estimate is to make a judgment about the size or the number of something without measuring or counting. This is important when the number or size of an object or population is too large or too difficult to accurately count or measure.

**Sample** Scientists may use a sample or a portion of the total number as a type of estimation. To sample is to take a small, representative portion of the objects or organisms of a population for research. By making careful observations or manipulating variables within that portion of the group, information is discovered and conclusions are drawn that might apply to the whole population. A poorly chosen sample can be unrepresentative of the whole. If you were trying to determine the rainfall in an area, it would not be best to take a rainfall sample from under a tree.

**Measure** You use measurements every day. Scientists also take measurements when collecting data. When taking measurements, it is important to know how to use measuring tools properly. Accuracy also is important.

**Length** To measure length, the distance between two points, scientists use meters. Smaller measurements might be measured in centimeters or millimeters.

Length is measured using a metric ruler or meterstick. When using a metric ruler, line up the 0-cm mark with the end of the object being measured and read the number of the unit where the object ends. Look at the metric ruler shown in **Figure 11.** The centimeter lines are the long, numbered lines, and the shorter lines are millimeter lines. In this instance, the length would be 4.50 cm.

**Figure 11** This metric ruler has centimeter and millimeter divisions.

SCIENCE SKILL HANDBOOK

MATH SKILL HANDBOOK

FOLDABLES HANDBOOK

REFERENCE HANDBOOK

GLOSSARY/ GLOSARIO

INDEX

SCIENCE SKILL HANDBOOK

MATH SKILL HANDBOOK

FOLDABLES HANDBOOK

REFERENCE HANDBOOK

GLOSSARY/ GLOSARIO

INDEX

**Mass** The SI unit for mass is the kilogram (kg). Scientists can measure mass using units formed by adding metric prefixes to the unit gram (g), such as milligram (mg). To measure mass, you might use a triple-beam balance similar to the one shown in **Figure 12.** The balance has a pan on one side and a set of beams on the other side. Each beam has a rider that slides on the beam.

When using a triple-beam balance, place an object on the pan. Slide the larg-est rider along its beam until the pointer drops below zero. Then move it back one notch. Repeat the process for each rider proceeding from the larger to smaller until the pointer swings an equal distance above and below the zero point. Sum the masses on each beam to find the mass of the object. Move all riders back to zero when finished.

Instead of putting materials directly on the balance, scientists often take a tare of a container. A tare is the mass of a container into which objects or substances are placed for measuring their masses. To find the mass of objects or substances, find the mass of a clean container. Remove the container from the pan, and place the object or sub-stances in the container. Find the mass of the container with the materials in it. Sub-tract the mass of the empty container from the mass of the filled container to find the mass of the materials you are using.

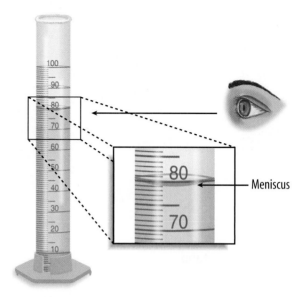

**Figure 13** Graduated cylinders measure liquid volume.

**Liquid Volume** To measure liquids, the unit used is the liter. When a smaller unit is needed, scientists might use a milliliter. Because a milliliter takes up the volume of a cube measuring 1 cm on each side it also can be called a cubic centimeter ($cm^3 = cm \times cm \times cm$).

You can use beakers and graduated cylinders to measure liquid volume. A graduated cylinder, shown in **Figure 13,** is marked from bottom to top in milliliters. In lab, you might use a 10-mL graduated cylinder or a 100-mL graduated cylinder. When measuring liquids, notice that the liquid has a curved surface. Look at the surface at eye level, and measure the bot-tom of the curve. This is called the menis-cus. The graduated cylinder in **Figure 13** contains 79.0 mL, or 79.0 $cm^3$, of a liquid.

**Temperature** Scientists often measure tem-perature using the Celsius scale. Pure water has a freezing point of 0°C and boiling point of 100°C. The unit of measurement is degrees Celsius. Two other scales often used are the Fahrenheit and Kelvin scales.

**Figure 12** A triple-beam balance is used to determine the mass of an object.

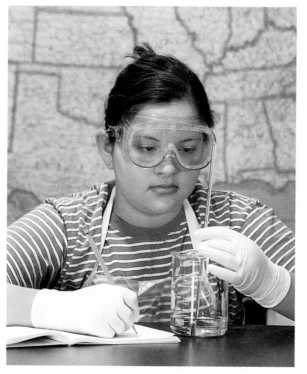

**Figure 14** A thermometer measures the temperature of an object.

Scientists use a thermometer to measure temperature. Most thermometers in a laboratory are glass tubes with a bulb at the bottom end containing a liquid such as colored alcohol. The liquid rises or falls with a change in temperature. To read a glass thermometer like the thermometer in **Figure 14,** rotate it slowly until a red line appears. Read the temperature where the red line ends.

### Form Operational Definitions
An operational definition defines an object by how it functions, works, or behaves. For example, when you are playing hide and seek and a tree is home base, you have created an operational definition for a tree.

Objects can have more than one operational definition. For example, a ruler can be defined as a tool that measures the length of an object (how it is used). It can also be a tool with a series of marks used as a standard when measuring (how it works).

## Analyze the Data

To determine the meaning of your observations and investigation results, you will need to look for patterns in the data. Then you must think critically to determine what the data mean. Scientists use several approaches when they analyze the data they have collected and recorded. Each approach is useful for identifying specific patterns.

### Interpret Data
The word *interpret* means "to explain the meaning of something." When analyzing data from an experiment, try to find out what the data show. Identify the control group and the test group to see whether changes in the independent variable have had an effect. Look for differences in the dependent variable between the control and test groups.

### Classify
Sorting objects or events into groups based on common features is called classifying. When classifying, first observe the objects or events to be classified. Then select one feature that is shared by some members in the group, but not by all. Place those members that share that feature in a subgroup. You can classify members into smaller and smaller subgroups based on characteristics. Remember that when you classify, you are grouping objects or events for a purpose. Keep your purpose in mind as you select the features to form groups and subgroups.

### Compare and Contrast
Observations can be analyzed by noting the similarities and differences between two or more objects or events that you observe. When you look at objects or events to see how they are similar, you are comparing them. Contrasting is looking for differences in objects or events.

SCIENCE SKILL HANDBOOK

MATH SKILL HANDBOOK

FOLDABLES HANDBOOK

REFERENCE HANDBOOK

GLOSSARY/ GLOSARIO

INDEX

Science Skill Handbook

Math Skill Handbook

Foldables Handbook

Reference Handbook

Glossary/ Glosario

Index

**Recognize Cause and Effect** A cause is a reason for an action or condition. The effect is that action or condition. When two events happen together, it is not necessarily true that one event caused the other. Scientists must design a controlled investigation to recognize the exact cause and effect.

## Draw Conclusions

When scientists have analyzed the data they collected, they proceed to draw conclusions about the data. These conclusions are sometimes stated in words similar to the hypothesis that you formed earlier. They may confirm a hypothesis, or lead you to a new hypothesis.

**Infer** Scientists often make inferences based on their observations. An inference is an attempt to explain observations or to indicate a cause. An inference is not a fact, but a logical conclusion that needs further investigation. For example, you may infer that a fire has caused smoke. Until you investigate, however, you do not know for sure.

**Apply** When you draw a conclusion, you must apply those conclusions to determine whether the data supports the hypothesis. If your data do not support your hypothesis, it does not mean that the hypothesis is wrong. It means only that the result of the investigation did not support the hypothesis. Maybe the experiment needs to be redesigned, or some of the initial observations on which the hypothesis was based were incomplete or biased. Perhaps more observation or research is needed to refine your hypothesis. A successful investigation does not always come out the way you originally predicted.

**Avoid Bias** Sometimes a scientific investigation involves making judgments. When you make a judgment, you form an opinion. It is important to be honest and not to allow any expectations of results to bias your judgments. This is important throughout the entire investigation, from researching to collecting data to drawing conclusions.

## Communicate

The communication of ideas is an important part of the work of scientists. A discovery that is not reported will not advance the scientific community's understanding or knowledge. Communication among scientists also is important as a way of improving their investigations.

Scientists communicate in many ways, from writing articles in journals and magazines that explain their investigations and experiments, to announcing important discoveries on television and radio. Scientists also share ideas with colleagues on the Internet or present them as lectures, like the student is doing in **Figure 15.**

**Figure 15** A student communicates to his peers about his investigation.

These safety symbols are used in laboratory and field investigations in this book to indicate possible hazards. Learn the meaning of each symbol and refer to this page often. *Remember to wash your hands thoroughly after completing lab procedures.*

## PROTECTIVE EQUIPMENT   Do not begin any lab without the proper protection equipment.

| | | | | | | | |
|---|---|---|---|---|---|---|---|
| **GOGGLES**  | Proper eye protection must be worn when performing or observing science activities that involve items or conditions as listed below. | **APRON**  | Wear an approved apron when using substances that could stain, wet, or destroy cloth. | **SOAP**  | Wash hands with soap and water before removing goggles and after all lab activities. | **GLOVES**  | Wear gloves when working with biological materials, chemicals, animals, or materials that can stain or irritate hands. |

## LABORATORY HAZARDS

| Symbols | Potential Hazards | Precaution | Response |
|---|---|---|---|
| **DISPOSAL** | contamination of classroom or environment due to improper disposal of materials such as chemicals and live specimens | • DO NOT dispose of hazardous materials in the sink or trash can.<br>• Dispose of wastes as directed by your teacher. | • If hazardous materials are disposed of improperly, notify your teacher immediately. |
| **EXTREME TEMPERATURE** | skin burns due to extremely hot or cold materials such as hot glass, liquids, or metals; liquid nitrogen; dry ice | • Use proper protective equipment, such as hot mitts and/or tongs, when handling objects with extreme temperatures. | • If injury occurs, notify your teacher immediately. |
| **SHARP OBJECTS** | punctures or cuts from sharp objects such as razor blades, pins, scalpels, and broken glass | • Handle glassware carefully to avoid breakage.<br>• Walk with sharp objects pointed downward, away from you and others. | • If broken glass or injury occurs, notify your teacher immediately. |
| **ELECTRICAL** | electric shock or skin burn due to improper grounding, short circuits, liquid spills, or exposed wires | • Check condition of wires and apparatus for fraying or uninsulated wires, and broken or cracked equipment.<br>• Use only GFCI-protected outlets | • DO NOT attempt to fix electrical problems. Notify your teacher immediately. |
| **CHEMICAL** | skin irritation or burns, breathing difficulty, and/or poisoning due to touching, swallowing, or inhalation of chemicals such as acids, bases, bleach, metal compounds, iodine, poinsettias, pollen, ammonia, acetone, nail polish remover, heated chemicals, mothballs, and any other chemicals labeled or known to be dangerous | • Wear proper protective equipment such as goggles, apron, and gloves when using chemicals.<br>• Ensure proper room ventilation or use a fume hood when using materials that produce fumes.<br>• NEVER smell fumes directly.<br>• NEVER taste or eat any material in the laboratory. | • If contact occurs, immediately flush affected area with water and notify your teacher.<br>• If a spill occurs, leave the area immediately and notify your teacher. |
| **FLAMMABLE** | unexpected fire due to liquids or gases that ignite easily such as rubbing alcohol | • Avoid open flames, sparks, or heat when flammable liquids are present. | • If a fire occurs, leave the area immediately and notify your teacher. |
| **OPEN FLAME** | burns or fire due to open flame from matches, Bunsen burners, or burning materials | • Tie back loose hair and clothing.<br>• Keep flame away from all materials.<br>• Follow teacher instructions when lighting and extinguishing flames.<br>• Use proper protection, such as hot mitts or tongs, when handling hot objects. | • If a fire occurs, leave the area immediately and notify your teacher. |
| **ANIMAL SAFETY** | injury to or from laboratory animals | • Wear proper protective equipment such as gloves, apron, and goggles when working with animals.<br>• Wash hands after handling animals. | • If injury occurs, notify your teacher immediately. |
| **BIOLOGICAL** | infection or adverse reaction due to contact with organisms such as bacteria, fungi, and biological materials such as blood, animal or plant materials | • Wear proper protective equipment such as gloves, goggles, and apron when working with biological materials.<br>• Avoid skin contact with an organism or any part of the organism.<br>• Wash hands after handling organisms. | • If contact occurs, wash the affected area and notify your teacher immediately. |
| **FUME** | breathing difficulties from inhalation of fumes from substances such as ammonia, acetone, nail polish remover, heated chemicals, and mothballs | • Wear goggles, apron, and gloves.<br>• Ensure proper room ventilation or use a fume hood when using substances that produce fumes.<br>• NEVER smell fumes directly. | • If a spill occurs, leave area and notify your teacher immediately. |
| **IRRITANT** | irritation of skin, mucous membranes, or respiratory tract due to materials such as acids, bases, bleach, pollen, mothballs, steel wool, and potassium permanganate | • Wear goggles, apron, and gloves.<br>• Wear a dust mask to protect against fine particles. | • If skin contact occurs, immediately flush the affected area with water and notify your teacher. |
| **RADIOACTIVE** | excessive exposure from alpha, beta, and gamma particles | • Remove gloves and wash hands with soap and water before removing remainder of protective equipment. | • If cracks or holes are found in the container, notify your teacher immediately. |

# Safety in the Science Laboratory

Science Skill Handbook

Math Skill Handbook

Foldables Handbook

Reference Handbook

Glossary/ Glosario

Index

## Introduction to Science Safety

The science laboratory is a safe place to work if you follow standard safety procedures. Being responsible for your own safety helps to make the entire laboratory a safer place for everyone. When performing any lab, read and apply the caution statements and safety symbol listed at the beginning of the lab.

## General Safety Rules

1. Complete the *Lab Safety Form* or other safety contract BEFORE starting any science lab.

2. Study the procedure. Ask your teacher any questions. Be sure you understand safety symbols shown on the page.

3. Notify your teacher about allergies or other health conditions that can affect your participation in a lab.

4. Learn and follow use and safety procedures for your equipment. If unsure, ask your teacher.

5. Never eat, drink, chew gum, apply cosmetics, or do any personal grooming in the lab. Never use lab glassware as food or drink containers. Keep your hands away from your face and mouth.

6. Know the location and proper use of the safety shower, eye wash, fire blanket, and fire alarm.

## Prevent Accidents

1. Use the safety equipment provided to you. Goggles and a safety apron should be worn during investigations.

2. Do NOT use hair spray, mousse, or other flammable hair products. Tie back long hair and tie down loose clothing.

3. Do NOT wear sandals or other open-toed shoes in the lab.

4. Remove jewelry on hands and wrists. Loose jewelry, such as chains and long necklaces, should be removed to prevent them from getting caught in equipment.

5. Do not taste any substances or draw any material into a tube with your mouth.

6. Proper behavior is expected in the lab. Practical jokes and fooling around can lead to accidents and injury.

7. Keep your work area uncluttered.

## Laboratory Work

1. Collect and carry all equipment and materials to your work area before beginning a lab.

2. Remain in your own work area unless given permission by your teacher to leave it.

3. Always slant test tubes away from yourself and others when heating them, adding substances to them, or rinsing them.

4. If instructed to smell a substance in a container, hold the container a short distance away and fan vapors toward your nose.

5. Do NOT substitute other chemicals/substances for those in the materials list unless instructed to do so by your teacher.

6. Do NOT take any materials or chemicals outside of the laboratory.

7. Stay out of storage areas unless instructed to be there and supervised by your teacher.

## Laboratory Cleanup

1. Turn off all burners, water, and gas, and disconnect all electrical devices.

2. Clean all pieces of equipment and return all materials to their proper places.

3. Dispose of chemicals and other materials as directed by your teacher. Place broken glass and solid substances in the proper containers. Never discard materials in the sink.

4. Clean your work area.

5. Wash your hands with soap and water thoroughly BEFORE removing your goggles.

## Emergencies

1. Report any fire, electrical shock, glassware breakage, spill, or injury, no matter how small, to your teacher immediately. Follow his or her instructions.

2. If your clothing should catch fire, STOP, DROP, and ROLL. If possible, smother it with the fire blanket or get under a safety shower. NEVER RUN.

3. If a fire should occur, turn off all gas and leave the room according to established procedures.

4. In most instances, your teacher will clean up spills. Do NOT attempt to clean up spills unless you are given permission and instructions to do so.

5. If chemicals come into contact with your eyes or skin, notify your teacher immediately. Use the eyewash, or flush your skin or eyes with large quantities of water.

6. The fire extinguisher and first-aid kit should only be used by your teacher unless it is an extreme emergency and you have been given permission.

7. If someone is injured or becomes ill, only a professional medical provider or someone certified in first aid should perform first-aid procedures.

SCIENCE SKILL HANDBOOK

MATH SKILL HANDBOOK

FOLDABLES HANDBOOK

REFERENCE HANDBOOK

GLOSSARY/ GLOSARIO

INDEX

SCIENCE SKILL HANDBOOK

MATH SKILL HANDBOOK

FOLDABLES HANDBOOK

REFERENCE HANDBOOK

GLOSSARY/ GLOSARIO

INDEX

## Use Fractions

A fraction compares a part to a whole. In the fraction $\frac{2}{3}$, the 2 represents the part and is the numerator. The 3 represents the whole and is the denominator.

**Reduce Fractions** To reduce a fraction, you must find the largest factor that is common to both the numerator and the denominator, the greatest common factor (GCF). Divide both numbers by the GCF. The fraction has then been reduced, or it is in its simplest form.

### Example

Twelve of the 20 chemicals in the science lab are in powder form. What fraction of the chemicals used in the lab are in powder form?

**Step 1** Write the fraction.

$$\frac{part}{whole} = \frac{12}{20}$$

**Step 2** To find the GCF of the numerator and denominator, list all of the factors of each number.

Factors of 12: 1, 2, 3, 4, 6, 12 (the numbers that divide evenly into 12)

Factors of 20: 1, 2, 4, 5, 10, 20 (the numbers that divide evenly into 20)

**Step 3** List the common factors.

1, 2, 4

**Step 4** Choose the greatest factor in the list. The GCF of 12 and 20 is 4.

**Step 5** Divide the numerator and denominator by the GCF.

$$\frac{12 \div 4}{20 \div 4} = \frac{3}{5}$$

In the lab, $\frac{3}{5}$ of the chemicals are in powder form.

**Practice Problem** At an amusement park, 66 of 90 rides have a height restriction. What fraction of the rides, in its simplest form, has a height restriction?

**Add and Subtract Fractions with Like Denominators** To add or subtract fractions with the same denominator, add or subtract the numerators and write the sum or difference over the denominator. After finding the sum or difference, find the simplest form for your fraction.

### Example 1

In the forest outside your house, $\frac{1}{8}$ of the animals are rabbits, $\frac{3}{8}$ are squirrels, and the remainder are birds and insects. How many are mammals?

**Step 1** Add the numerators.

$$\frac{1}{8} + \frac{3}{8} = \frac{(1 + 3)}{8} = \frac{4}{8}$$

**Step 2** Find the GCF.

$$\frac{4}{8} \text{ (GCF, 4)}$$

**Step 3** Divide the numerator and denominator by the GCF.

$$\frac{4 \div 4}{8 \div 4} = \frac{1}{2}$$

$\frac{1}{2}$ of the animals are mammals.

### Example 2

If $\frac{7}{16}$ of the Earth is covered by freshwater, and $\frac{1}{16}$ of that is in glaciers, how much freshwater is not frozen?

**Step 1** Subtract the numerators.

$$\frac{7}{16} - \frac{1}{16} = \frac{(7 - 1)}{16} = \frac{6}{16}$$

**Step 2** Find the GCF.

$$\frac{6}{16} \text{ (GCF, 2)}$$

**Step 3** Divide the numerator and denominator by the GCF.

$$\frac{6 \div 2}{16 \div 2} = \frac{3}{8}$$

$\frac{3}{8}$ of the freshwater is not frozen.

**Practice Problem** A bicycle rider is riding at a rate of 15 km/h for $\frac{4}{9}$ of his ride, 10 km/h for $\frac{2}{9}$ of his ride, and 8 km/h for the remainder of the ride. How much of his ride is he riding at a rate greater than 8 km/h?

**Add and Subtract Fractions with Unlike Denominators** To add or subtract fractions with unlike denominators, first find the least common denominator (LCD). This is the smallest number that is a common multiple of both denominators. Rename each fraction with the LCD, and then add or subtract. Find the simplest form if necessary.

### Example 1

A chemist makes a paste that is $\frac{1}{2}$ table salt (NaCl), $\frac{1}{3}$ sugar ($C_6H_{12}O_6$), and the remainder is water ($H_2O$). How much of the paste is a solid?

**Step 1** Find the LCD of the fractions.

$\frac{1}{2} + \frac{1}{3}$ (LCD, 6)

**Step 2** Rename each numerator and each denominator with the LCD.

**Step 3** Add the numerators.

$\frac{3}{6} + \frac{2}{6} = \frac{(3+2)}{6} = \frac{5}{6}$

$\frac{5}{6}$ of the paste is a solid.

### Example 2

The average precipitation in Grand Junction, CO, is $\frac{7}{10}$ inch in November, and $\frac{3}{5}$ inch in December. What is the total average precipitation?

**Step 1** Find the LCD of the fractions.

$\frac{7}{10} + \frac{3}{5}$ (LCD, 10)

**Step 2** Rename each numerator and each denominator with the LCD.

**Step 3** Add the numerators.

$\frac{7}{10} + \frac{6}{10} = \frac{(7+6)}{10} = \frac{13}{10}$

$\frac{13}{10}$ inches total precipitation, or $1\frac{3}{10}$ inches.

**Practice Problem** On an electric bill, about $\frac{1}{8}$ of the energy is from solar energy and about $\frac{1}{10}$ is from wind power. How much of the total bill is from solar energy and wind power combined?

### Example 3

In your body, $\frac{7}{10}$ of your muscle contractions are involuntary (cardiac and smooth muscle tissue). Smooth muscle makes $\frac{3}{15}$ of your muscle contractions. How many of your muscle contractions are made by cardiac muscle?

**Step 1** Find the LCD of the fractions.

$\frac{7}{10} - \frac{3}{15}$ (LCD, 30)

**Step 2** Rename each numerator and each denominator with the LCD.

$\frac{7 \times 3}{10 \times 3} = \frac{21}{30}$

$\frac{3 \times 2}{15 \times 2} = \frac{6}{30}$

**Step 3** Subtract the numerators.

$\frac{21}{30} - \frac{6}{30} = \frac{(21-6)}{30} = \frac{15}{30}$

**Step 4** Find the GCF.

$\frac{15}{30}$ (GCF, 15)

$\frac{1}{2}$

$\frac{1}{2}$ of all muscle contractions are cardiac muscle.

### Example 4

Tony wants to make cookies that call for $\frac{3}{4}$ of a cup of flour, but he only has $\frac{1}{3}$ of a cup. How much more flour does he need?

**Step 1** Find the LCD of the fractions.

$\frac{3}{4} - \frac{1}{3}$ (LCD, 12)

**Step 2** Rename each numerator and each denominator with the LCD.

$\frac{3 \times 3}{4 \times 3} = \frac{9}{12}$

$\frac{1 \times 4}{3 \times 4} = \frac{4}{12}$

**Step 3** Subtract the numerators.

$\frac{9}{12} - \frac{4}{12} = \frac{(9-4)}{12} = \frac{5}{12}$

$\frac{5}{12}$ of a cup of flour

**Practice Problem** Using the information provided to you in Example 3 above, determine how many muscle contractions are voluntary (skeletal muscle).

SCIENCE SKILL HANDBOOK

MATH SKILL HANDBOOK

FOLDABLES HANDBOOK

REFERENCE HANDBOOK

GLOSSARY/ GLOSARIO

INDEX

SCIENCE SKILL HANDBOOK

MATH SKILL HANDBOOK

FOLDABLES HANDBOOK

REFERENCE HANDBOOK

GLOSSARY/ GLOSARIO

INDEX

**Multiply Fractions** To multiply with fractions, multiply the numerators and multiply the denominators. Find the simplest form if necessary.

### Example

Multiply $\frac{3}{5}$ by $\frac{1}{3}$.

**Step 1**   Multiply the numerators and denominators.

$$\frac{3}{5} \times \frac{1}{3} = \frac{(3 \times 1)}{(5 \times 3)} \frac{3}{15}$$

**Step 2**   Find the GCF.

$$\frac{3}{15} \text{ (GCF, 3)}$$

**Step 3**   Divide the numerator and denominator by the GCF.

$$\frac{3 \div 3}{15 \div 3} = \frac{1}{5}$$

$\frac{3}{5}$ multiplied by $\frac{1}{3}$ is $\frac{1}{5}$.

**Practice Problem**  Multiply $\frac{3}{14}$ by $\frac{5}{16}$.

**Find a Reciprocal** Two numbers whose product is 1 are called multiplicative inverses, or reciprocals.

### Example

Find the reciprocal of $\frac{3}{8}$.

**Step 1**   Inverse the fraction by putting the denominator on top and the numerator on the bottom.

$$\frac{8}{3}$$

The reciprocal of $\frac{3}{8}$ is $\frac{8}{3}$.

**Practice Problem**  Find the reciprocal of $\frac{4}{9}$.

**Divide Fractions** To divide one fraction by another fraction, multiply the dividend by the reciprocal of the divisor. Find the simplest form if necessary.

### Example 1

Divide $\frac{1}{9}$ by $\frac{1}{3}$.

**Step 1**   Find the reciprocal of the divisor. The reciprocal of $\frac{1}{3}$ is $\frac{3}{1}$.

**Step 2**   Multiply the dividend by the reciprocal of the divisor.

$$\frac{\frac{1}{9}}{\frac{1}{3}} = \frac{1}{9} \times \frac{3}{1} = \frac{(1 \times 3)}{(9 \times 1)} = \frac{3}{9}$$

**Step 3**   Find the GCF.

$$\frac{3}{9} \text{ (GCF, 3)}$$

**Step 4**   Divide the numerator and denominator by the GCF.

$$\frac{3 \div 3}{9 \div 3} = \frac{1}{3}$$

$\frac{1}{9}$ divided by $\frac{1}{3}$ is $\frac{1}{3}$.

### Example 2

Divide $\frac{3}{5}$ by $\frac{1}{4}$.

**Step 1**   Find the reciprocal of the divisor. The reciprocal of $\frac{1}{4}$ is $\frac{4}{1}$.

**Step 2**   Multiply the dividend by the reciprocal of the divisor.

$$\frac{\frac{3}{5}}{\frac{1}{4}} = \frac{3}{5} \times \frac{4}{1} = \frac{(3 \times 4)}{(5 \times 1)} = \frac{12}{5}$$

$\frac{3}{5}$ divided by $\frac{1}{4}$ is $\frac{12}{5}$ or $2\frac{2}{5}$.

**Practice Problem**  Divide $\frac{3}{11}$ by $\frac{7}{10}$.

## Use Ratios

When you compare two numbers by division, you are using a ratio. Ratios can be written 3 to 5, 3:5, or $\frac{3}{5}$. Ratios, like fractions, also can be written in simplest form.

Ratios can represent one type of probability, called odds. This is a ratio that compares the number of ways a certain outcome occurs to the number of possible outcomes. For example, if you flip a coin 100 times, what are the odds that it will come up heads? There are two possible outcomes, heads or tails, so the odds of coming up heads are 50:100. Another way to say this is that 50 out of 100 times the coin will come up heads. In its simplest form, the ratio is 1:2.

### Example 1

A chemical solution contains 40 g of salt and 64 g of baking soda. What is the ratio of salt to baking soda as a fraction in simplest form?

**Step 1**    Write the ratio as a fraction.

$$\frac{\text{salt}}{\text{baking soda}} = \frac{40}{64}$$

**Step 2**    Express the fraction in simplest form. The GCF of 40 and 64 is 8.

$$\frac{40}{64} = \frac{40 \div 8}{64 \div 8} = \frac{5}{8}$$

The ratio of salt to baking soda in the sample is 5:8.

### Example 2

Sean rolls a 6-sided die 6 times. What are the odds that the side with a 3 will show?

**Step 1**    Write the ratio as a fraction.

$$\frac{\text{number of sides with a 3}}{\text{number of possible sides}} = \frac{1}{6}$$

**Step 2**    Multiply by the number of attempts.

$$\frac{1}{6} \times 6 \text{ attempts} = \frac{6}{6} \text{ attempts} = 1 \text{ attempt}$$

1 attempt out of 6 will show a 3.

**Practice Problem**  Two metal rods measure 100 cm and 144 cm in length. What is the ratio of their lengths in simplest form?

## Use Decimals

A fraction with a denominator that is a power of ten can be written as a decimal. For example, 0.27 means $\frac{27}{100}$. The decimal point separates the ones place from the tenths place.

Any fraction can be written as a decimal using division. For example, the fraction $\frac{5}{8}$ can be written as a decimal by dividing 5 by 8. Written as a decimal, it is 0.625.

**Add or Subtract Decimals**  When adding and subtracting decimals, line up the decimal points before carrying out the operation.

### Example 1

Find the sum of 47.68 and 7.80.

**Step 1**    Line up the decimal places when you write the numbers.

$$\begin{array}{r} 47.68 \\ + \ 7.80 \\ \hline \end{array}$$

**Step 2**    Add the decimals.

$$\begin{array}{r} {}^{1\ 1}\phantom{0} \\ 47.68 \\ + \ 7.80 \\ \hline 55.48 \end{array}$$

The sum of 47.68 and 7.80 is 55.48.

### Example 2

Find the difference of 42.17 and 15.85.

**Step 1**    Line up the decimal places when you write the number.

$$\begin{array}{r} 42.17 \\ -15.85 \\ \hline \end{array}$$

**Step 2**    Subtract the decimals.

$$\begin{array}{r} {}^{3\ 11}\phantom{0} \\ 42.17 \\ -15.85 \\ \hline 26.32 \end{array}$$

The difference of 42.17 and 15.85 is 26.32.

**Practice Problem**  Find the sum of 1.245 and 3.842.

SCIENCE SKILL HANDBOOK

MATH SKILL HANDBOOK

FOLDABLES HANDBOOK

REFERENCE HANDBOOK

GLOSSARY/ GLOSARIO

INDEX

**Multiply Decimals** To multiply decimals, multiply the numbers like numbers without decimal points. Count the decimal places in each factor. The product will have the same number of decimal places as the sum of the decimal places in the factors.

### Example

Multiply 2.4 by 5.9.

**Step 1** Multiply the factors like two whole numbers.

$24 \times 59 = 1416$

**Step 2** Find the sum of the number of decimal places in the factors. Each factor has one decimal place, for a sum of two decimal places.

**Step 3** The product will have two decimal places.

14.16

The product of 2.4 and 5.9 is 14.16.

**Practice Problem** Multiply 4.6 by 2.2.

**Divide Decimals** When dividing decimals, change the divisor to a whole number. To do this, multiply both the divisor and the dividend by the same power of ten. Then place the decimal point in the quotient directly above the decimal point in the dividend. Then divide as you do with whole numbers.

### Example

Divide 8.84 by 3.4.

**Step 1** Multiply both factors by 10.

$3.4 \times 10 = 34, 8.84 \times 10 = 88.4$

**Step 2** Divide 88.4 by 34.

```
        2.6
   34)88.4
      −68
       204
      −204
         0
```

8.84 divided by 3.4 is 2.6.

**Practice Problem** Divide 75.6 by 3.6.

## Use Proportions

An equation that shows that two ratios are equivalent is a proportion. The ratios $\frac{2}{4}$ and $\frac{5}{10}$ are equivalent, so they can be written as $\frac{2}{4} = \frac{5}{10}$. This equation is a proportion.

When two ratios form a proportion, the cross products are equal. To find the cross products in the proportion $\frac{2}{4} = \frac{5}{10}$, multiply the 2 and the 10, and the 4 and the 5. Therefore $2 \times 10 = 4 \times 5$, or $20 = 20$.

Because you know that both ratios are equal, you can use cross products to find a missing term in a proportion. This is known as solving the proportion.

### Example

The heights of a tree and a pole are proportional to the lengths of their shadows. The tree casts a shadow of 24 m when a 6-m pole casts a shadow of 4 m. What is the height of the tree?

**Step 1** Write a proportion.

$$\frac{\text{height of tree}}{\text{height of pole}} = \frac{\text{length of tree's shadow}}{\text{length of pole's shadow}}$$

**Step 2** Substitute the known values into the proportion. Let $h$ represent the unknown value, the height of the tree.

$$\frac{h}{6} \times \frac{24}{4}$$

**Step 3** Find the cross products.

$h \times 4 = 6 \times 24$

**Step 4** Simplify the equation.

$4h \times 144$

**Step 5** Divide each side by 4.

$$\frac{4h}{4} \times \frac{144}{4}$$

$h = 36$

The height of the tree is 36 m.

**Practice Problem** The ratios of the weights of two objects on the Moon and on Earth are in proportion. A rock weighing 3 N on the Moon weighs 18 N on Earth. How much would a rock that weighs 5 N on the Moon weigh on Earth?

## Use Percentages

The word *percent* means "out of one hundred." It is a ratio that compares a number to 100. Suppose you read that 77 percent of Earth's surface is covered by water. That is the same as reading that the fraction of Earth's surface covered by water is $\frac{77}{100}$. To express a fraction as a percent, first find the equivalent decimal for the fraction. Then, multiply the decimal by 100 and add the percent symbol.

### Example 1

Express $\frac{13}{20}$ as a percent.

**Step 1** Find the equivalent decimal for the fraction.

$$\begin{array}{r} 0.65 \\ 20\overline{)13.00} \\ \underline{12\ 0} \\ 1\ 00 \\ \underline{1\ 00} \\ 0 \end{array}$$

**Step 2** Rewrite the fraction $\frac{13}{20}$ as 0.65.

**Step 3** Multiply 0.65 by 100 and add the % symbol.

$$0.65 \times 100 = 65 = 65\%$$

So, $\frac{13}{20} = 65\%$.

This also can be solved as a proportion.

### Example 2

Express $\frac{13}{20}$ as a percent.

**Step 1** Write a proportion.

$$\frac{13}{20} = \frac{x}{100}$$

**Step 2** Find the cross products.

$$1300 = 20x$$

**Step 3** Divide each side by 20.

$$\frac{1300}{20} = \frac{20x}{20}$$

$$65\% = x$$

**Practice Problem** In one year, 73 of 365 days were rainy in one city. What percent of the days in that city were rainy?

## Solve One-Step Equations

A statement that two expressions are equal is an equation. For example, $A = B$ is an equation that states that $A$ is equal to $B$.

An equation is solved when a variable is replaced with a value that makes both sides of the equation equal. To make both sides equal the inverse operation is used. Addition and subtraction are inverses, and multiplication and division are inverses.

### Example 1

Solve the equation $x - 10 = 35$.

**Step 1** Find the solution by adding 10 to each side of the equation.

$$x - 10 = 35$$
$$x - 10 + 10 = 35 - 10$$
$$x = 45$$

**Step 2** Check the solution.

$$x - 10 = 35$$
$$45 - 10 = 35$$
$$35 = 35$$

Both sides of the equation are equal, so $x = 45$.

### Example 2

In the formula $a = bc$, find the value of $c$ if $a = 20$ and $b = 2$.

**Step 1** Rearrange the formula so the unknown value is by itself on one side of the equation by dividing both sides by $b$.

$$a = bc$$
$$\frac{a}{b} = \frac{bc}{b}$$
$$\frac{a}{b} = c$$

**Step 2** Replace the variables $a$ and $b$ with the values that are given.

$$\frac{a}{b} = c$$
$$\frac{20}{2} = c$$
$$10 = c$$

**Step 3** Check the solution.

$$a = bc$$
$$20 = 2 \times 10$$
$$20 = 20$$

Both sides of the equation are equal, so $c = 10$ is the solution when $a = 20$ and $b = 2$.

**Practice Problem** In the formula $h = gd$, find the value of $d$ if $g = 12.3$ and $h = 17.4$.

SCIENCE SKILL HANDBOOK

MATH SKILL HANDBOOK

FOLDABLES HANDBOOK

REFERENCE HANDBOOK

GLOSSARY/ GLOSARIO

INDEX

SCIENCE SKILL HANDBOOK

MATH SKILL HANDBOOK

FOLDABLES HANDBOOK

REFERENCE HANDBOOK

GLOSSARY/GLOSARIO

INDEX

# Use Statistics

The branch of mathematics that deals with collecting, analyzing, and presenting data is statistics. In statistics, there are three common ways to summarize data with a single number—the mean, the median, and the mode.

The **mean** of a set of data is the arithmetic average. It is found by adding the numbers in the data set and dividing by the number of items in the set.

The **median** is the middle number in a set of data when the data are arranged in numerical order. If there were an even number of data points, the median would be the mean of the two middle numbers.

The **mode** of a set of data is the number or item that appears most often.

Another number that often is used to describe a set of data is the range. The **range** is the difference between the largest number and the smallest number in a set of data.

## Example

The speeds (in m/s) for a race car during five different time trials are 39, 37, 44, 36, and 44.

**To find the mean:**

**Step 1**  Find the sum of the numbers.

$39 + 37 + 44 + 36 + 44 = 200$

**Step 2**  Divide the sum by the number of items, which is 5.

$200 \div 5 = 40$

The mean is 40 m/s.

**To find the median:**

**Step 1**  Arrange the measures from least to greatest.

36, 37, 39, 44, 44

**Step 2**  Determine the middle measure.

36, 37, <u>39</u>, 44, 44

The median is 39 m/s.

**To find the mode:**

**Step 1**  Group the numbers that are the same together.

44, 44, 36, 37, 39

**Step 2**  Determine the number that occurs most in the set.

<u>44, 44</u>, 36, 37, 39

The mode is 44 m/s.

**To find the range:**

**Step 1**  Arrange the measures from greatest to least.

44, 44, 39, 37, 36

**Step 2**  Determine the greatest and least measures in the set.

<u>44</u>, 44, 39, 37, <u>36</u>

**Step 3**  Find the difference between the greatest and least measures.

$44 - 36 = 8$

The range is 8 m/s.

**Practice Problem**  Find the mean, median, mode, and range for the data set 8, 4, 12, 8, 11, 14, 16.

A **frequency table** shows how many times each piece of data occurs, usually in a survey. **Table 1** below shows the results of a student survey on favorite color.

| Table 1  Student Color Choice | | |
|---|---|---|
| **Color** | **Tally** | **Frequency** |
| red | IIII | 4 |
| blue | IͶ̄L | 5 |
| black | II | 2 |
| green | III | 3 |
| purple | IͶ̄L II | 7 |
| yellow | IͶ̄L I | 6 |

Based on the frequency table data, which color is the favorite?

## Use Geometry

The branch of mathematics that deals with the measurement, properties, and relationships of points, lines, angles, surfaces, and solids is called geometry.

**Perimeter** The **perimeter** ($P$) is the distance around a geometric figure. To find the perimeter of a rectangle, add the length and width and multiply that sum by two, or $2(l + w)$. To find perimeters of irregular figures, add the length of the sides.

### Example 1

Find the perimeter of a rectangle that is 3 m long and 5 m wide.

**Step 1**  You know that the perimeter is 2 times the sum of the width and length.

$P = 2(3 \text{ m} + 5 \text{ m})$

**Step 2**  Find the sum of the width and length.

$P = 2(8 \text{ m})$

**Step 3**  Multiply by 2.

$P = 16 \text{ m}$

The perimeter is 16 m.

### Example 2

Find the perimeter of a shape with sides measuring 2 cm, 5 cm, 6 cm, 3 cm.

**Step 1**  You know that the perimeter is the sum of all the sides.

$P = 2 + 5 + 6 + 3$

**Step 2**  Find the sum of the sides.

$P = 2 + 5 + 6 + 3$

$P = 16$

The perimeter is 16 cm.

**Practice Problem**  Find the perimeter of a rectangle with a length of 18 m and a width of 7 m.

**Practice Problem**  Find the perimeter of a triangle measuring 1.6 cm by 2.4 cm by 2.4 cm.

**Area of a Rectangle**  The **area** ($A$) is the number of square units needed to cover a surface. To find the area of a rectangle, multiply the length times the width, or $l \times w$. When finding area, the units also are multiplied. Area is given in square units.

### Example

Find the area of a rectangle with a length of 1 cm and a width of 10 cm.

**Step 1**  You know that the area is the length multiplied by the width.

$A = (1 \text{ cm} \times 10 \text{ cm})$

**Step 2**  Multiply the length by the width. Also multiply the units.

$A = 10 \text{ cm}^2$

The area is 10 cm².

**Practice Problem**  Find the area of a square whose sides measure 4 m.

**Area of a Triangle**  To find the area of a triangle, use the formula:

$A = \frac{1}{2}(\text{base} \times \text{height})$

The base of a triangle can be any of its sides. The height is the perpendicular distance from a base to the opposite endpoint, or vertex.

### Example

Find the area of a triangle with a base of 18 m and a height of 7 m.

**Step 1**  You know that the area is $\frac{1}{2}$ the base times the height.

$A = \frac{1}{2}(18 \text{ m} \times 7 \text{ m})$

**Step 2**  Multiply $\frac{1}{2}$ by the product of $18 \times 7$. Multiply the units.

$A = \frac{1}{2}(126 \text{ m}^2)$

$A = 63 \text{ m}^2$

The area is 63 m².

**Practice Problem**  Find the area of a triangle with a base of 27 cm and a height of 17 cm.

SCIENCE SKILL HANDBOOK

MATH SKILL HANDBOOK

FOLDABLES HANDBOOK

REFERENCE HANDBOOK

GLOSSARY/ GLOSARIO

INDEX

SCIENCE SKILL HANDBOOK

MATH SKILL HANDBOOK

FOLDABLES HANDBOOK

REFERENCE HANDBOOK

GLOSSARY/ GLOSARIO

INDEX

**Circumference of a Circle** The **diameter** (*d*) of a circle is the distance across the circle through its center, and the **radius** (r) is the distance from the center to any point on the circle. The radius is half of the diameter. The distance around the circle is called the **circumference** (C). The formula for finding the circumference is:

$$C = 2\pi r \text{ or } C = \pi d$$

The circumference divided by the diameter is always equal to 3.1415926… This nonterminating and nonrepeating number is represented by the Greek letter $\pi$ (pi). An approximation often used for $\pi$ is 3.14.

### Example 1

Find the circumference of a circle with a radius of 3 m.

**Step 1** You know the formula for the circumference is 2 times the radius times $\pi$.

$$C = 2\pi(3)$$

**Step 2** Multiply 2 times the radius.

$$C = 6\pi$$

**Step 3** Multiply by $\pi$.

$$C \approx 19 \text{ m}$$

The circumference is about 19 m.

### Example 2

Find the circumference of a circle with a diameter of 24.0 cm.

**Step 1** You know the formula for the circumference is the diameter times $\pi$.

$$C = \pi(24.0)$$

**Step 2** Multiply the diameter by $\pi$.

$$C \approx 75.4 \text{ cm}$$

The circumference is about 75.4 cm.

**Practice Problem** Find the circumference of a circle with a radius of 19 cm.

**Area of a Circle** The formula for the area of a circle is: $A = \pi r^2$

### Example 1

Find the area of a circle with a radius of 4.0 cm.

**Step 1** $A = \pi(4.0)^2$

**Step 2** Find the square of the radius.

$$A = 16\pi$$

**Step 3** Multiply the square of the radius by $\pi$.

$$A \approx 50 \text{ cm}^2$$

The area of the circle is about 50 cm².

### Example 2

Find the area of a circle with a radius of 225 m.

**Step 1** $A = \pi(225)^2$

**Step 2** Find the square of the radius.

$$A = 50625\pi$$

**Step 3** Multiply the square of the radius by $\pi$.

$$A \approx 159043.1$$

The area of the circle is about 159043.1 m².

### Example 3

Find the area of a circle whose diameter is 20.0 mm.

**Step 1** Remember that the radius is half of the diameter.

$$A = \pi\left(\frac{20.0}{2}\right)^2$$

**Step 2** Find the radius.

$$A = \pi(10.0)^2$$

**Step 3** Find the square of the radius.

$$A = 100\pi$$

**Step 4** Multiply the square of the radius by $\pi$.

$$A \approx 314 \text{ mm}^2$$

The area of the circle is about 314 mm².

**Practice Problem** Find the area of a circle with a radius of 16 m.

**Volume** The measure of space occupied by a solid is the **volume** ($V$). To find the volume of a rectangular solid multiply the length times width times height, or $V = l \times w \times h$. It is measured in cubic units, such as cubic centimeters ($cm^3$).

**Example**

Find the volume of a rectangular solid with a length of 2.0 m, a width of 4.0 m, and a height of 3.0 m.

**Step 1** You know the formula for volume is the length times the width times the height.

$$V = 2.0 \text{ m} \times 4.0 \text{ m} \times 3.0 \text{ m}$$

**Step 2** Multiply the length times the width times the height.

$$V = 24 \text{ m}^3$$

The volume is 24 m$^3$.

**Practice Problem** Find the volume of a rectangular solid that is 8 m long, 4 m wide, and 4 m high.

To find the volume of other solids, multiply the area of the base times the height.

**Example 1**

Find the volume of a solid that has a triangular base with a length of 8.0 m and a height of 7.0 m. The height of the entire solid is 15.0 m.

**Step 1** You know that the base is a triangle, and the area of a triangle is $\frac{1}{2}$ the base times the height, and the volume is the area of the base times the height.

$$V = \left[\frac{1}{2}(b \times h)\right] \times 15$$

**Step 2** Find the area of the base.

$$V = \left[\frac{1}{2}(8 \times 7)\right] \times 15$$
$$V = \left(\frac{1}{2} \times 56\right) \times 15$$

**Step 3** Multiply the area of the base by the height of the solid.

$$V = 28 \times 15$$
$$V = 420 \text{ m}^3$$

The volume is 420 m$^3$.

**Example 2**

Find the volume of a cylinder that has a base with a radius of 12.0 cm, and a height of 21.0 cm.

**Step 1** You know that the base is a circle, and the area of a circle is the square of the radius times $\pi$, and the volume is the area of the base times the height.

$$V = (\pi r^2) \times 21$$
$$V = (\pi 12^2) \times 21$$

**Step 2** Find the area of the base.

$$V = 144\pi \times 21$$
$$V = 452 \times 21$$

**Step 3** Multiply the area of the base by the height of the solid.

$$V \approx 9{,}500 \text{ cm}^3$$

The volume is about 9,500 cm$^3$.

**Example 3**

Find the volume of a cylinder that has a diameter of 15 mm and a height of 4.8 mm.

**Step 1** You know that the base is a circle with an area equal to the square of the radius times $\pi$. The radius is one-half the diameter. The volume is the area of the base times the height.

$$V = (\pi r^2) \times 4.8$$
$$V = \left[\pi\left(\frac{1}{2} \times 15\right)^2\right] \times 4.8$$
$$V = (\pi 7.5^2) \times 4.8$$

**Step 2** Find the area of the base.

$$V = 56.25\pi \times 4.8$$
$$V \approx 176.71 \times 4.8$$

**Step 3** Multiply the area of the base by the height of the solid.

$$V \approx 848.2$$

The volume is about 848.2 mm$^3$.

**Practice Problem** Find the volume of a cylinder with a diameter of 7 cm in the base and a height of 16 cm.

SCIENCE SKILL HANDBOOK

MATH SKILL HANDBOOK

FOLDABLES HANDBOOK

REFERENCE HANDBOOK

GLOSSARY/ GLOSARIO

INDEX

# Science Applications

SCIENCE SKILL HANDBOOK

MATH SKILL HANDBOOK

FOLDABLES HANDBOOK

REFERENCE HANDBOOK

GLOSSARY/ GLOSARIO

INDEX

## Measure in SI

The metric system of measurement was developed in 1795. A modern form of the metric system, called the International System (SI), was adopted in 1960 and provides the standard measurements that all scientists around the world can understand.

The SI system is convenient because unit sizes vary by powers of 10. Prefixes are used to name units. Look at **Table 2** for some common SI prefixes and their meanings.

| Table 2 | Common SI Prefixes | | |
|---------|--------|--------|-----------|
| **Prefix** | **Symbol** | **Meaning** | |
| *kilo–* | k | 1,000 | thousandth |
| *hecto–* | h | 100 | hundred |
| *deka–* | da | 10 | ten |
| *deci–* | d | 0.1 | tenth |
| *centi–* | c | 0.01 | hundreth |
| *milli–* | m | 0.001 | thousandth |

### Example

How many grams equal one kilogram?

**Step 1**  Find the prefix *kilo–* in **Table 2.**

**Step 2**  Using **Table 2,** determine the meaning of *kilo–*. According to the table, it means 1,000. When the prefix *kilo–* is added to a unit, it means that there are 1,000 of the units in a "kilounit."

**Step 3**  Apply the prefix to the units in the question. The units in the question are grams. There are 1,000 grams in a kilogram.

**Practice Problem**  Is a milligram larger or smaller than a gram? How many of the smaller units equal one larger unit? What fraction of the larger unit does one smaller unit represent?

## Dimensional Analysis

**Convert SI Units**  In science, quantities such as length, mass, and time sometimes are measured using different units. A process called dimensional analysis can be used to change one unit of measure to another. This process involves multiplying your starting quantity and units by one or more conversion factors. A conversion factor is a ratio equal to one and can be made from any two equal quantities with different units. If 1,000 mL equal 1 L then two ratios can be made.

$$\frac{1,000 \text{ mL}}{1 \text{ L}} = \frac{1 \text{ L}}{1,000 \text{ mL}} = 1$$

One can convert between units in the SI system by using the equivalents in **Table 2** to make conversion factors.

### Example

How many cm are in 4 m?

**Step 1**  Write conversion factors for the units given. From **Table 2,** you know that 100 cm = 1 m. The conversion factors are

$$\frac{100 \text{ cm}}{1 \text{ m}} \text{ and } \frac{1 \text{ m}}{100 \text{ cm}}$$

**Step 2**  Decide which conversion factor to use. Select the factor that has the units you are converting from (m) in the denominator and the units you are converting to (cm) in the numerator.

$$\frac{100 \text{ cm}}{1 \text{ m}}$$

**Step 3**  Multiply the starting quantity and units by the conversion factor. Cancel the starting units with the units in the denominator. There are 400 cm in 4 m.

$$4 \text{ m} = \frac{100 \text{ cm}}{1 \text{ m}} = 400 \text{ cm}$$

**Practice Problem**  How many milligrams are in one kilogram? (Hint: You will need to use two conversion factors from **Table 2.**)

## Table 3  Unit System Equivalents

| Type of Measurement | Equivalent |
|---|---|
| Length | 1 in = 2.54 cm<br>1 yd = 0.91 m<br>1 mi = 1.61 km |
| Mass and weight* | 1 oz = 28.35 g<br>1 lb = 0.45 kg<br>1 ton (short) = 0.91 tonnes (metric tons)<br>1 lb = 4.45 N |
| Volume | $1 \text{ in}^3 = 16.39 \text{ cm}^3$<br>1 qt = 0.95 L<br>1 gal = 3.78 L |
| Area | $1 \text{ in}^2 = 6.45 \text{ cm}^2$<br>$1 \text{ yd}^2 = 0.83 \text{ m}^2$<br>$1 \text{ mi}^2 = 2.59 \text{ km}^2$<br>1 acre = 0.40 hectares |
| Temperature | $°C = \dfrac{(°F - 32)}{1.8}$<br>$K = °C + 273$ |

*Weight is measured in standard Earth gravity.

**Convert Between Unit Systems**  **Table 3** gives a list of equivalents that can be used to convert between English and SI units.

### Example

If a meterstick has a length of 100 cm, how long is the meterstick in inches?

**Step 1**  Write the conversion factors for the units given. From **Table 3,** 1 in = 2.54 cm.

$$\frac{1 \text{ in}}{2.54 \text{ cm}} \quad and \quad \frac{2.54 \text{ cm}}{1 \text{ in}}$$

**Step 2**  Determine which conversion factor to use. You are converting from cm to in. Use the conversion factor with cm on the bottom.

$$\frac{1 \text{ in}}{2.54 \text{ cm}}$$

**Step 3**  Multiply the starting quantity and units by the conversion factor. Cancel the starting units with the units in the denominator. Round your answer to the nearest tenth.

$$100 \text{ cm} \times \frac{1 \text{ in}}{2.54 \text{ cm}} = 39.37 \text{ in}$$

The meterstick is about 39.4 in long.

**Practice Problem 1**  A book has a mass of 5 lb. What is the mass of the book in kg?

**Practice Problem 2**  Use the equivalent for in and cm (1 in = 2.54 cm) to show how $1 \text{ in}^3 \approx 16.39 \text{ cm}^3$.

SCIENCE SKILL HANDBOOK

MATH SKILL HANDBOOK

FOLDABLES HANDBOOK

REFERENCE HANDBOOK

GLOSSARY/ GLOSARIO

INDEX

SCIENCE SKILL HANDBOOK

MATH SKILL HANDBOOK

FOLDABLES HANDBOOK

REFERENCE HANDBOOK

GLOSSARY/ GLOSARIO

INDEX

## Precision and Significant Digits

When you make a measurement, the value you record depends on the precision of the measuring instrument. This precision is represented by the number of significant digits recorded in the measurement. When counting the number of significant digits, all digits are counted except zeros at the end of a number with no decimal point such as 2,050, and zeros at the beginning of a decimal such as 0.03020. When adding or subtracting numbers with different precision, round the answer to the smallest number of decimal places of any number in the sum or difference. When multiplying or dividing, the answer is rounded to the smallest number of significant digits of any number being multiplied or divided.

### Example

The lengths 5.28 and 5.2 are measured in meters. Find the sum of these lengths and record your answer using the correct number of significant digits.

**Step 1**   Find the sum.

      5.28 m    2 digits after the decimal
+  5.2 m    1 digit after the decimal
  10.48 m

**Step 2**   Round to one digit after the decimal because the least number of digits after the decimal of the numbers being added is 1.

The sum is 10.5 m.

**Practice Problem 1**   How many significant digits are in the measurement 7,071,301 m? How many significant digits are in the measurement 0.003010 g?

**Practice Problem 2**   Multiply 5.28 and 5.2 using the rule for multiplying and dividing. Record the answer using the correct number of significant digits.

## Scientific Notation

Many times numbers used in science are very small or very large. Because these numbers are difficult to work with scientists use scientific notation. To write numbers in scientific notation, move the decimal point until only one non-zero digit remains on the left. Then count the number of places you moved the decimal point and use that number as a power of ten. For example, the average distance from the Sun to Mars is 227,800,000,000 m. In scientific notation, this distance is $2.278 \times 10^{11}$ m. Because you moved the decimal point to the left, the number is a positive power of ten.

The mass of an electron is about 0.000 000 000 000 000 000 000 000 000 000 911 kg. Expressed in scientific notation, this mass is $9.11 \times 10^{-31}$ kg. Because the decimal point was moved to the right, the number is a negative power of ten.

### Example

Earth is 149,600,000 km from the Sun. Express this in scientific notation.

**Step 1**   Move the decimal point until one non-zero digit remains on the left.

    1.496 000 00

**Step 2**   Count the number of decimal places you have moved. In this case, eight.

**Step 2**   Show that number as a power of ten, $10^8$.

Earth is $1.496 \times 10^8$ km from the Sun.

**Practice Problem 1**   How many significant digits are in 149,600,000 km? How many significant digits are in $1.496 \times 10^8$ km?

**Practice Problem 2**   Parts used in a high performance car must be measured to $7 \times 10^{-6}$ m. Express this number as a decimal.

**Practice Problem 3**   A CD is spinning at 539 revolutions per minute. Express this number in scientific notation.

## Make and Use Graphs

Data in tables can be displayed in a graph—a visual representation of data. Common graph types include line graphs, bar graphs, and circle graphs.

**Line Graph** A line graph shows a relationship between two variables that change continuously. The independent variable is changed and is plotted on the x-axis. The dependent variable is observed, and is plotted on the y-axis.

### Example

Draw a line graph of the data below from a cyclist in a long-distance race.

| Table 4 Bicycle Race Data | |
|---|---|
| Time (h) | Distance (km) |
| 0 | 0 |
| 1 | 8 |
| 2 | 16 |
| 3 | 24 |
| 4 | 32 |
| 5 | 40 |

**Step 1** Determine the x-axis and y-axis variables. Time varies independently of distance and is plotted on the x-axis. Distance is dependent on time and is plotted on the y-axis.

**Step 2** Determine the scale of each axis. The x-axis data ranges from 0 to 5. The y-axis data ranges from 0 to 50.

**Step 3** Using graph paper, draw and label the axes. Include units in the labels.

**Step 4** Draw a point at the intersection of the time value on the x-axis and corresponding distance value on the y-axis. Connect the points and label the graph with a title, as shown in **Figure 8.**

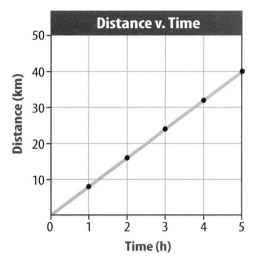

**Figure 8** This line graph shows the relationship between distance and time during a bicycle ride.

**Practice Problem** A puppy's shoulder height is measured during the first year of her life. The following measurements were collected: (3 mo, 52 cm), (6 mo, 72 cm), (9 mo, 83 cm), (12 mo, 86 cm). Graph this data.

**Find a Slope** The slope of a straight line is the ratio of the vertical change, rise, to the horizontal change, run.

$$\text{Slope} = \frac{\text{vertical change (rise)}}{\text{horizontal change (run)}} = \frac{\text{change in } y}{\text{change in } x}$$

### Example

Find the slope of the graph in **Figure 8**.

**Step 1** You know that the slope is the change in y divided by the change in x.

$$\text{Slope} = \frac{\text{change in } y}{\text{change in } x}$$

**Step 2** Determine the data points you will be using. For a straight line, choose the two sets of points that are the farthest apart.

$$\text{Slope} = \frac{(40 - 0) \text{ km}}{(5 - 0) \text{ h}}$$

**Step 3** Find the change in y and x.

$$\text{Slope} = \frac{40 \text{ km}}{5 \text{ h}}$$

**Step 4** Divide the change in y by the change in x.

$$\text{Slope} = \frac{8 \text{ km}}{\text{h}}$$

The slope of the graph is 8 km/h.

SCIENCE SKILL HANDBOOK

MATH SKILL HANDBOOK

FOLDABLES HANDBOOK

REFERENCE HANDBOOK

GLOSSARY/ GLOSARIO

INDEX

SCIENCE SKILL HANDBOOK

MATH SKILL HANDBOOK

FOLDABLES HANDBOOK

REFERENCE HANDBOOK

GLOSSARY/ GLOSARIO

INDEX

**Bar Graph** To compare data that does not change continuously you might choose a bar graph. A bar graph uses bars to show the relationships between variables. The *x*-axis variable is divided into parts. The parts can be numbers such as years, or a category such as a type of animal. The *y*-axis is a number and increases continuously along the axis.

### Example

A recycling center collects 4.0 kg of aluminum on Monday, 1.0 kg on Wednesday, and 2.0 kg on Friday. Create a bar graph of this data.

**Step 1** Select the *x*-axis and *y*-axis variables. The measured numbers (the masses of aluminum) should be placed on the *y*-axis. The variable divided into parts (collection days) is placed on the *x*-axis.

**Step 2** Create a graph grid like you would for a line graph. Include labels and units.

**Step 3** For each measured number, draw a vertical bar above the *x*-axis value up to the *y*-axis value. For the first data point, draw a vertical bar above Monday up to 4.0 kg.

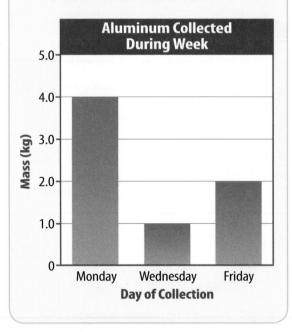

**Practice Problem** Draw a bar graph of the gases in air: 78% nitrogen, 21% oxygen, 1% other gases.

**Circle Graph** To display data as parts of a whole, you might use a circle graph. A circle graph is a circle divided into sections that represent the relative size of each piece of data. The entire circle represents 100%, half represents 50%, and so on.

### Example

Air is made up of 78% nitrogen, 21% oxygen, and 1% other gases. Display the composition of air in a circle graph.

**Step 1** Multiply each percent by 360° and divide by 100 to find the angle of each section in the circle.

$$78\% \times \frac{360°}{100} = 280.8°$$

$$21\% \times \frac{360°}{100} = 75.6°$$

$$1\% \times \frac{360°}{100} = 3.6°$$

**Step 2** Use a compass to draw a circle and to mark the center of the circle. Draw a straight line from the center to the edge of the circle.

**Step 3** Use a protractor and the angles you calculated to divide the circle into parts. Place the center of the protractor over the center of the circle and line the base of the protractor over the straight line.

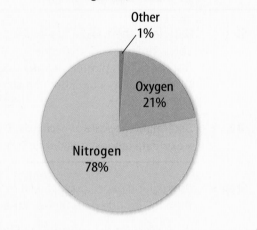

**Practice Problem** Draw a circle graph to represent the amount of aluminum collected during the week shown in the bar graph to the left.

# FOLDABLES® Handbook

## Student Study Guides & Instructions
### By Dinah Zike

1. You will find suggestions for Study Guides, also known as Foldables or books, in each chapter lesson and as a final project. Look at the end of the chapter to determine the project format and glue the Foldables in place as you progress through the chapter lessons.

2. Creating the Foldables or books is simple and easy to do by using copy paper, art paper, and internet printouts. Photocopies of maps, diagrams, or your own illustrations may also be used for some of the Foldables. Notebook paper is the most common source of material for study guides and 83% of all Foldables are created from it. When folded to make books, notebook paper Foldables easily fit into 11″ × 17″ or 12″ × 18″ chapter projects with space left over. Foldables made using photocopy paper are slightly larger and they fit into Projects, but snugly. Use the least amount of glue, tape, and staples needed to assemble the Foldables.

3. Seven of the Foldables can be made using either small or large paper. When 11″ × 17″ or 12″ × 18″ paper is used, these become projects for housing smaller Foldables. Project format boxes are located within the instructions to remind you of this option.

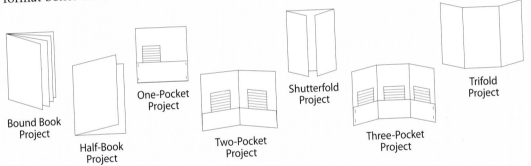

Bound Book Project

Half-Book Project

One-Pocket Project

Two-Pocket Project

Shutterfold Project

Three-Pocket Project

Trifold Project

4. Use one-gallon self-locking plastic bags to store your projects. Place strips of two-inch clear tape along the left, long side of the bag and punch holes through the taped edge. Cut the bottom corners off the bag so it will not hold air. Store this Project Portfolio inside a three-hole binder. To store a large collection of project bags, use a giant laundry-soap box. Holes can be punched in some of the Foldable Projects so they can be stored in a three-hole binder without using a plastic bag. Punch holes in the pocket books before gluing or stapling the pocket.

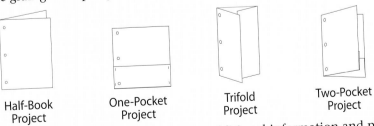

Half-Book Project

One-Pocket Project

Trifold Project

Two-Pocket Project

5. Maximize the use of the projects by collecting additional information and placing it on the back of the project and other unused spaces of the large Foldables.

SCIENCE SKILL HANDBOOK

MATH SKILL HANDBOOK

FOLDABLES HANDBOOK

REFERENCE HANDBOOK

GLOSSARY/ GLOSARIO

INDEX

# Half-Book Foldable® By Dinah Zike

**Step 1** Fold a sheet of notebook or copy paper in half.

Label the exterior tab and use the inside space to write information.

### PROJECT FORMAT
Use 11" × 17" or 12" × 18" paper on the horizontal axis to make a large project book.

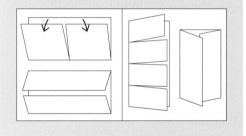

## Variations

Paper can be folded horizontally, like a *hamburger* or vertically, like a *hot dog*.

**A**

**B**

**C** Half-books can be folded so that one side is ½ inch longer than the other side. A title or question can be written on the extended tab.

---

# Worksheet Foldable or Folded Book® By Dinah Zike

**Step 1** Make a half-book (see above) using work sheets, internet print-outs, diagrams, or maps.

**Step 2** Fold it in half again.

## Variations

**A** This folded sheet as a small book with two pages can be used for comparing and contrasting, cause and effect, or other skills.

**B** When the sheet of paper is open, the four sections can be used separately or used collectively to show sequences or steps.

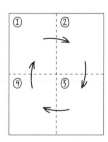

SCIENCE SKILL HANDBOOK

MATH SKILL HANDBOOK

FOLDABLES HANDBOOK

REFERENCE HANDBOOK

GLOSSARY/ GLOSARIO

INDEX

# Two-Tab and Concept-Map Foldable® By Dinah Zike

**Step 1** Fold a sheet of notebook or copy paper in half vertically or horizontally.

**Step 2** Fold it in half again, as shown.

**Step 3** Unfold once and cut along the fold line or valley of the top flap to make two flaps.

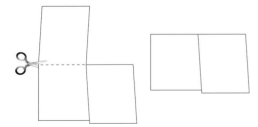

## Variations

**A** Concept maps can be made by leaving a ½ inch tab at the top when folding the paper in half. Use arrows and labels to relate topics to the primary concept.

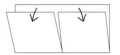

**B** Use two sheets of paper to make multiple page tab books. Glue or staple books together at the top fold.

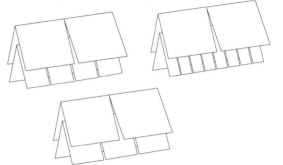

- - - - - - - - - - - - - - - - - - - - - - - - - - - - - - - - - - - - - - - -

# Three-Quarter Foldable® By Dinah Zike

**Step 1** Make a two-tab book (see above) and cut the left tab off at the top of the fold line.

## Variations

**A** Use this book to draw a diagram or a map on the exposed left tab. Write questions about the illustration on the top right tab and provide complete answers on the space under the tab.

**B** Compose a self-test using multiple choice answers for your questions. Include the correct answer with three wrong responses. The correct answers can be written on the back of the book or upside down on the bottom of the inside page.

SCIENCE SKILL HANDBOOK

MATH SKILL HANDBOOK

FOLDABLES HANDBOOK

REFERENCE HANDBOOK

GLOSSARY/ GLOSARIO

INDEX

Science Skill Handbook

Math Skill Handbook

Foldables Handbook

Reference Handbook

Glossary/ Glosario

Index

# Three-Tab Foldable® By Dinah Zike

**Step 1** Fold a sheet of paper in half horizontally.

**Step 2** Fold into thirds.

**Step 3** Unfold and cut along the folds of the top flap to make three sections.

## Variations

**A** Before cutting the three tabs draw a Venn diagram across the front of the book.

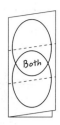

**B** Make a space to use for titles or concept maps by leaving a ½ inch tab at the top when folding the paper in half.

# Four-Tab Foldable® By Dinah Zike

**Step 1** Fold a sheet of paper in half horizontally.

**Step 2** Fold in half and then fold each half as shown below.

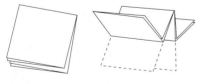

**Step 3** Unfold and cut along the fold lines of the top flap to make four tabs.

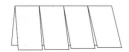

## Variations

**A** Make a space to use for titles or concept maps by leaving a ½ inch tab at the top when folding the paper in half.

**B** Use the book on the vertical axis, with or without an extended tab.

# Folding Fifths for a Foldable® By Dinah Zike

**Step 1** Fold a sheet of paper in half horizontally.

**Step 2** Fold again so one-third of the paper is exposed and two-thirds are covered.

**Step 3** Fold the two-thirds section in half.

**Step 4** Fold the one-third section, a single thickness, backward to make a fold line.

## Variations

**A** Unfold and cut along the fold lines to make five tabs.

**B** Make a five-tab book with a ½ inch tab at the top (see two-tab instructions).

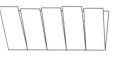

**C** Use 11″ × 17″ or 12″ × 18″ paper and fold into fifths for a five-column and/or row table or chart.

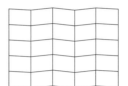

---

# Folded Table or Chart, and Trifold Foldable® By Dinah Zike

**Step 1** Fold a sheet of paper in the required number of vertical columns for the table or chart.

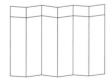

**Step 2** Fold the horizontal rows needed for the table or chart.

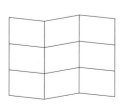

## Variations

**A** Make a trifold by folding the paper into thirds vertically or horizontally.

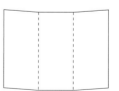

**B** Make a trifold book. Unfold it and draw a Venn diagram on the inside.

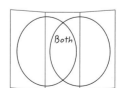

**PROJECT FORMAT**
Use 11″ × 17″ or 12″ × 18″ paper and fold it to make a large trifold project book or larger tables and charts.

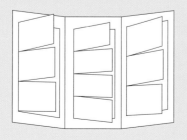

SCIENCE SKILL HANDBOOK

MATH SKILL HANDBOOK

FOLDABLES HANDBOOK

REFERENCE HANDBOOK

GLOSSARY/ GLOSARIO

INDEX

# Two or Three-Pockets Foldable® By Dinah Zike

**Step 1** Fold up the long side of a horizontal sheet of paper about 5 cm.

**Step 2** Fold the paper in half.

**Step 3** Open the paper and glue or staple the outer edges to make two compartments.

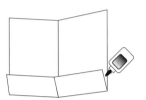

## Variations

**A** Make a multi-page booklet by gluing several pocket books together.

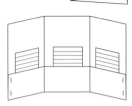

**B** Make a three-pocket book by using a trifold (see previous instructions).

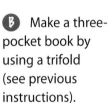

**PROJECT FORMAT**

Use 11″ × 17″ or 12″ × 18″ paper and fold it horizontally to make a large multi-pocket project.

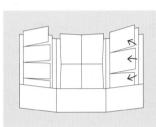

---

# Matchbook Foldable® By Dinah Zike

**Step 1** Fold a sheet of paper almost in half and make the back edge about 1–2 cm longer than the front edge.

**Step 2** Find the midpoint of the shorter flap.

**Step 3** Open the paper and cut the short side along the midpoint making two tabs.

**Step 4** Close the book and fold the tab over the short side.

## Variations

**A** Make a single-tab matchbook by skipping Steps 2 and 3.

**B** Make two smaller matchbooks by cutting the single-tab matchbook in half.

SCIENCE SKILL HANDBOOK

MATH SKILL HANDBOOK

FOLDABLES HANDBOOK

REFERENCE HANDBOOK

GLOSSARY/ GLOSARIO

INDEX

# Shutterfold Foldable® By Dinah Zike

**Step 1** Begin as if you were folding a vertical sheet of paper in half, but instead of creasing the paper, pinch it to show the midpoint.

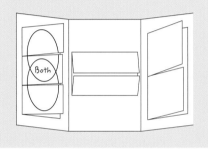

**PROJECT FORMAT**
Use 11" × 17" or 12" × 18" paper and fold it to make a large shutterfold project.

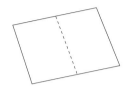

**Step 2** Fold the top and bottom to the middle and crease the folds.

## Variations

**A** Use the shutterfold on the horizontal axis.

**B** Create a center tab by leaving .5–2 cm between the flaps in Step 2.

# Four-Door Foldable® By Dinah Zike

**Step 1** Make a shutterfold (see above).

**Step 2** Fold the sheet of paper in half.

**Step 3** Open the last fold and cut along the inside fold lines to make four tabs.

## Variations

**A** Use the four-door book on the opposite axis.

**B** Create a center tab by leaving .5–2 cm between the flaps in Step 1.

SCIENCE SKILL HANDBOOK

MATH SKILL HANDBOOK

FOLDABLES HANDBOOK

REFERENCE HANDBOOK

GLOSSARY/GLOSARIO

INDEX

# Bound Book Foldable® By Dinah Zike

**Step 1** Fold three sheets of paper in half. Place the papers in a stack, leaving about .5 cm between each top fold. Mark all three sheets about 3 cm from the outer edges.

**Step 2** Using two of the sheets, cut from the outer edges to the marked spots on each side. On the other sheet, cut between the marked spots.

**Step 3** Take the two sheets from Step 1 and slide them through the cut in the third sheet to make a 12-page book.

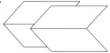

**Step 4** Fold the bound pages in half to form a book.

## Variation

**A** Use two sheets of paper to make an eight-page book, or increase the number of pages by using more than three sheets.

### PROJECT FORMAT
Use two or more sheets of 11″ × 17″ or 12″ × 18″ paper and fold it to make a large bound book project.

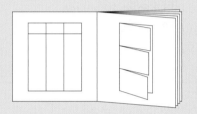

---

# Accordian Foldable® By Dinah Zike

**Step 1** Fold the selected paper in half vertically, like a *hamburger*.

**Step 2** Cut each sheet of folded paper in half along the fold lines.

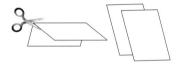

**Step 3** Fold each half-sheet almost in half, leaving a 2 cm tab at the top.

**Step 4** Fold the top tab over the short side, then fold it in the opposite direction.

## Variations

**A** Glue the straight edge of one paper inside the tab of another sheet. Leave a tab at the end of the book to add more pages.

**B** Tape the straight edge of one paper to the tab of another sheet, or just tape the straight edges of nonfolded paper end to end to make an accordian.

**C** Use whole sheets of paper to make a large accordian.

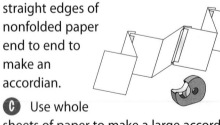

SCIENCE SKILL HANDBOOK

MATH SKILL HANDBOOK

FOLDABLES HANDBOOK

REFERENCE HANDBOOK

GLOSSARY/ GLOSARIO

INDEX

# Layered Foldable® By Dinah Zike

**Step 1** Stack two sheets of paper about 1–2 cm apart. Keep the right and left edges even.

**Step 2** Fold up the bottom edges to form four tabs. Crease the fold to hold the tabs in place.

**Step 3** Staple along the folded edge, or open and glue the papers together at the fold line.

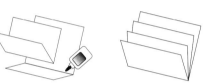

## Variations

**A** Rotate the book so the fold is at the top or to the side.

**B** Extend the book by using more than two sheets of paper.

# Envelope Foldable® By Dinah Zike

**Step 1** Fold a sheet of paper into a *taco*. Cut off the tab at the top.

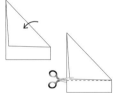

**Step 2** Open the *taco* and fold it the opposite way making another *taco* and an X-fold pattern on the sheet of paper.

**Step 3** Cut a map, illustration, or diagram to fit the inside of the envelope.

**Step 4** Use the outside tabs for labels and inside tabs for writing information.

## Variations

**A** Use 11″ × 17″ or 12″ × 18″ paper to make a large envelope.

**B** Cut off the points of the four tabs to make a window in the middle of the book.

SCIENCE SKILL HANDBOOK

MATH SKILL HANDBOOK

FOLDABLES HANDBOOK

REFERENCE HANDBOOK

GLOSSARY/ GLOSARIO

INDEX

# Sentence Strip Foldable® By Dinah Zike

**Step 1** Fold two sheets of paper in half vertically, like a *hamburger*.

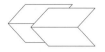

**Step 2** Unfold and cut along fold lines making four half sheets.

**Step 3** Fold each half sheet in half horizontally, like a *hot dog*.

**Step 4** Stack folded horizontal sheets evenly and staple together on the left side.

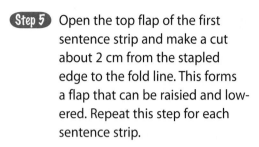

**Step 5** Open the top flap of the first sentence strip and make a cut about 2 cm from the stapled edge to the fold line. This forms a flap that can be raisied and lowered. Repeat this step for each sentence strip.

## Variations

**A** Expand this book by using more than two sheets of paper.

**B** Use whole sheets of paper to make large books.

# Pyramid Foldable® By Dinah Zike

**Step 1** Fold a sheet of paper into a *taco*. Crease the fold line, but do not cut it off.

**Step 2** Open the folded sheet and refold it like a *taco* in the opposite direction to create an X-fold pattern.

**Step 3** Cut one fold line as shown, stopping at the center of the X-fold to make a flap.

**Step 4** Outline the fold lines of the X-fold. Label the three front sections and use the inside spaces for notes. Use the tab for the title.

**Step 5** Glue the tab into a project book or notebook. Use the space under the pyramid for other information.

Title:

**Step 6** To display the pyramid, fold the flap under and secure with a paper clip, if needed.

Title:

SCIENCE SKILL HANDBOOK

MATH SKILL HANDBOOK

FOLDABLES HANDBOOK

REFERENCE HANDBOOK

GLOSSARY/ GLOSARIO

INDEX

# Single-Pocket or One-Pocket Foldable® By Dinah Zike

**Step 1** Using a large piece of paper on a vertical axis, fold the bottom edge of the paper upwards, about 5 cm.

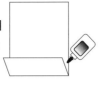

**Step 2** Glue or staple the outer edges to make a large pocket.

### PROJECT FORMAT
Use 11″ × 17″ or 12″ × 18″ paper and fold it vertically or horizontally to make a large pocket project.

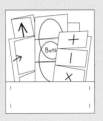

## Variations

**A** Make the one-pocket project using the paper on the horizontal axis.

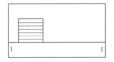

**B** To store materials securely inside, fold the top of the paper almost to the center, leaving about 2–4 cm between the paper edges. Slip the Foldables through the opening and under the top and bottom pockets.

---

# Multi-Tab Foldable® By Dinah Zike

**Step 1** Fold a sheet of notebook paper in half like a *hot dog*.

**Step 2** Open the paper and on one side cut every third line. This makes ten tabs on wide ruled notebook paper and twelve tabs on college ruled.

**Step 3** Label the tabs on the front side and use the inside space for definitions or other information.

## Variation

**A** Make a tab for a title by folding the paper so the holes remain uncovered. This allows the notebook Foldable to be stored in a three-hole binder.

SCIENCE SKILL HANDBOOK

MATH SKILL HANDBOOK

FOLDABLES HANDBOOK

REFERENCE HANDBOOK

GLOSSARY/ GLOSARIO

INDEX

# PERIODIC TABLE OF THE ELEMENTS

Element — Hydrogen
Atomic number — 1
Symbol — H
Atomic mass — 1.01
State of matter

Gas
Liquid
Solid
Synthetic

A column in the periodic table is called a **group**.

A row in the periodic table is called a **period**.

The number in parentheses is the mass number of the longest lived isotope for that element.

| Group | 1 | 2 | 3 | 4 | 5 | 6 | 7 | 8 | 9 |
|---|---|---|---|---|---|---|---|---|---|
| Period 1 | Hydrogen 1 H 1.01 | | | | | | | | |
| Period 2 | Lithium 3 Li 6.94 | Beryllium 4 Be 9.01 | | | | | | | |
| Period 3 | Sodium 11 Na 22.99 | Magnesium 12 Mg 24.31 | | | | | | | |
| Period 4 | Potassium 19 K 39.10 | Calcium 20 Ca 40.08 | Scandium 21 Sc 44.96 | Titanium 22 Ti 47.87 | Vanadium 23 V 50.94 | Chromium 24 Cr 52.00 | Manganese 25 Mn 54.94 | Iron 26 Fe 55.85 | Cobalt 27 Co 58.93 |
| Period 5 | Rubidium 37 Rb 85.47 | Strontium 38 Sr 87.62 | Yttrium 39 Y 88.91 | Zirconium 40 Zr 91.22 | Niobium 41 Nb 92.91 | Molybdenum 42 Mo 95.96 | Technetium 43 Tc (98) | Ruthenium 44 Ru 101.07 | Rhodium 45 Rh 102.91 |
| Period 6 | Cesium 55 Cs 132.91 | Barium 56 Ba 137.33 | Lanthanum 57 La 138.91 | Hafnium 72 Hf 178.49 | Tantalum 73 Ta 180.95 | Tungsten 74 W 183.84 | Rhenium 75 Re 186.21 | Osmium 76 Os 190.23 | Iridium 77 Ir 192.22 |
| Period 7 | Francium 87 Fr (223) | Radium 88 Ra (226) | Actinium 89 Ac (227) | Rutherfordium 104 Rf (267) | Dubnium 105 Db (268) | Seaborgium 106 Sg (271) | Bohrium 107 Bh (272) | Hassium 108 Hs (270) | Meitnerium 109 Mt (276) |

**Lanthanide series**

| Cerium 58 Ce 140.12 | Praseodymium 59 Pr 140.91 | Neodymium 60 Nd 144.24 | Promethium 61 Pm (145) | Samarium 62 Sm 150.36 | Europium 63 Eu 151.96 |
|---|---|---|---|---|---|

**Actinide series**

| Thorium 90 Th 232.04 | Protactinium 91 Pa 231.04 | Uranium 92 U 238.03 | Neptunium 93 Np (237) | Plutonium 94 Pu (244) | Americium 95 Am (243) |
|---|---|---|---|---|---|

Metal
Metalloid
Nonmetal
Recently discovered

**18**

| | | | | | | Helium<br>2<br>**He**<br>4.00 |

**13** **14** **15** **16** **17**

| Boron<br>5<br>**B**<br>10.81 | Carbon<br>6<br>**C**<br>12.01 | Nitrogen<br>7<br>**N**<br>14.01 | Oxygen<br>8<br>**O**<br>16.00 | Fluorine<br>9<br>**F**<br>19.00 | Neon<br>10<br>**Ne**<br>20.18 |
| Aluminum<br>13<br>**Al**<br>26.98 | Silicon<br>14<br>**Si**<br>28.09 | Phosphorus<br>15<br>**P**<br>30.97 | Sulfur<br>16<br>**S**<br>32.07 | Chlorine<br>17<br>**Cl**<br>35.45 | Argon<br>18<br>**Ar**<br>39.95 |

**10** **11** **12**

| Nickel<br>28<br>**Ni**<br>58.69 | Copper<br>29<br>**Cu**<br>63.55 | Zinc<br>30<br>**Zn**<br>65.38 | Gallium<br>31<br>**Ga**<br>69.72 | Germanium<br>32<br>**Ge**<br>72.64 | Arsenic<br>33<br>**As**<br>74.92 | Selenium<br>34<br>**Se**<br>78.96 | Bromine<br>35<br>**Br**<br>79.90 | Krypton<br>36<br>**Kr**<br>83.80 |
| Palladium<br>46<br>**Pd**<br>106.42 | Silver<br>47<br>**Ag**<br>107.87 | Cadmium<br>48<br>**Cd**<br>112.41 | Indium<br>49<br>**In**<br>114.82 | Tin<br>50<br>**Sn**<br>118.71 | Antimony<br>51<br>**Sb**<br>121.76 | Tellurium<br>52<br>**Te**<br>127.60 | Iodine<br>53<br>**I**<br>126.90 | Xenon<br>54<br>**Xe**<br>131.29 |
| Platinum<br>78<br>**Pt**<br>195.08 | Gold<br>79<br>**Au**<br>196.97 | Mercury<br>80<br>**Hg**<br>200.59 | Thallium<br>81<br>**Tl**<br>204.38 | Lead<br>82<br>**Pb**<br>207.20 | Bismuth<br>83<br>**Bi**<br>208.98 | Polonium<br>84<br>**Po**<br>(209) | Astatine<br>85<br>**At**<br>(210) | Radon<br>86<br>**Rn**<br>(222) |
| Darmstadtium<br>110<br>**Ds**<br>(281) | Roentgenium<br>111<br>**Rg**<br>(280) | Copernicium<br>112<br>**Cn**<br>(285) | * Ununtrium<br>113<br>**Uut**<br>(284) | * Ununquadium<br>114<br>**Uuq**<br>(289) | * Ununpentium<br>115<br>**Uup**<br>(288) | * Ununhexium<br>116<br>**Uuh**<br>(293) | | * Ununoctium<br>118<br>**Uuo**<br>(294) |

* The names and symbols for elements 113-116 and 118 are temporary. Final names will be selected when the elements' discoveries are verified.

| Gadolinium<br>64<br>**Gd**<br>157.25 | Terbium<br>65<br>**Tb**<br>158.93 | Dysprosium<br>66<br>**Dy**<br>162.50 | Holmium<br>67<br>**Ho**<br>164.93 | Erbium<br>68<br>**Er**<br>167.26 | Thulium<br>69<br>**Tm**<br>168.93 | Ytterbium<br>70<br>**Yb**<br>173.05 | Lutetium<br>71<br>**Lu**<br>174.97 |
| Curium<br>96<br>**Cm**<br>(247) | Berkelium<br>97<br>**Bk**<br>(247) | Californium<br>98<br>**Cf**<br>(251) | Einsteinium<br>99<br>**Es**<br>(252) | Fermium<br>100<br>**Fm**<br>(257) | Mendelevium<br>101<br>**Md**<br>(258) | Nobelium<br>102<br>**No**<br>(259) | Lawrencium<br>103<br>**Lr**<br>(262) |

SCIENCE SKILL HANDBOOK

MATH SKILL HANDBOOK

FOLDABLES HANDBOOK

REFERENCE HANDBOOK

GLOSSARY/GLOSARIO

INDEX

# Diversity of Life: Classification of Living Organisms

A six-kingdom system of classification of organisms is used today. Two kingdoms—Kingdom Archaebacteria and Kingdom Eubacteria—contain organisms that do not have a nucleus and that lack membrane-bound structures in the cytoplasm of their cells. The members of the other four kingdoms have a cell or cells that contain a nucleus and structures in the cytoplasm, some of which are surrounded by membranes. These kingdoms are Kingdom Protista, Kingdom Fungi, Kingdom Plantae, and Kingdom Animalia.

## Kingdom Archaebacteria

one-celled; some absorb food from their surroundings; some are photosynthetic; some are chemosynthetic; many are found in extremely harsh environments including salt ponds, hot springs, swamps, and deep-sea hydrothermal vents

## Kingdom Eubacteria

one-celled; most absorb food from their surroundings; some are photosynthetic; some are chemosynthetic; many are parasites; many are round, spiral, or rod-shaped; some form colonies

## Kingdom Protista

**Phylum Euglenophyta** one-celled; photosynthetic or take in food; most have one flagellum; euglenoids

**Kingdom Eubacteria**
Bacillus anthracis

**Phylum Chlorophyta**
Desmids

**Phylum Bacillariophyta** one-celled; photosynthetic; have unique double shells made of silica; diatoms

**Phylum Dinoflagellata** one-celled; photosynthetic; contain red pigments; have two flagella; dinoflagellates

**Phylum Chlorophyta** one-celled, many-celled, or colonies; photosynthetic; contain chlorophyll; live on land, in freshwater, or salt water; green algae

**Phylum Rhodophyta** most are many-celled; photosynthetic; contain red pigments; most live in deep, saltwater environments; red algae

**Phylum Phaeophyta** most are many-celled; photosynthetic; contain brown pigments; most live in saltwater environments; brown algae

**Phylum Rhizopoda** one-celled; take in food; are free-living or parasitic; move by means of pseudopods; amoebas

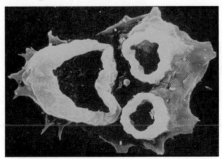

**Amoeba**

SCIENCE SKILL HANDBOOK

MATH SKILL HANDBOOK

FOLDABLES HANDBOOK

REFERENCE HANDBOOK

GLOSSARY/GLOSARIO

INDEX

**Phylum Zoomastigina** one-celled; take in food; free-living or parasitic; have one or more flagella; zoomastigotes

**Phylum Ciliophora** one-celled; take in food; have large numbers of cilia; ciliates

**Phylum Sporozoa** one-celled; take in food; have no means of movement; are parasites in animals; sporozoans

**Phylum Myxomycota**
Slime mold

**Phylum Oomycota**
Phytophthora infestans

**Phyla Myxomycota and Acrasiomycota** one- or many-celled; absorb food; change form during life cycle; cellular and plasmodial slime molds

**Phylum Oomycota** many-celled; are either parasites or decomposers; live in freshwater or salt water; water molds, rusts and downy mildews

## Kingdom Fungi

**Phylum Zygomycota** many-celled; absorb food; spores are produced in sporangia; zygote fungi; bread mold

**Phylum Ascomycota** one- and many-celled; absorb food; spores produced in asci; sac fungi; yeast

**Phylum Basidiomycota** many-celled; absorb food; spores produced in basidia; club fungi; mushrooms

**Phylum Deuteromycota** members with unknown reproductive structures; imperfect fungi; *Penicillium*

**Phylum Mycophycota** organisms formed by symbiotic relationship between an ascomycote or a basidiomycote and green alga or cyanobacterium; lichens

**Lichens**

SCIENCE SKILL HANDBOOK

MATH SKILL HANDBOOK

FOLDABLES HANDBOOK

REFERENCE HANDBOOK

GLOSSARY/ GLOSARIO

INDEX

## Kingdom Plantae

Divisions **Bryophyta** (mosses), **Anthocerophyta** (hornworts), **Hepaticophyta** (liverworts), **Psilophyta** (whisk ferns) many-celled non-vascular plants; reproduce by spores produced in capsules; green; grow in moist, land environments

Division **Lycophyta** many-celled vascular plants; spores are produced in conelike structures; live on land; are photosynthetic; club mosses

Division **Arthrophyta** vascular plants; ribbed and jointed stems; scalelike leaves; spores produced in conelike structures; horsetails

Division **Pterophyta** vascular plants; leaves called fronds; spores produced in clusters of sporangia called sori; live on land or in water; ferns

Division **Ginkgophyta** deciduous trees; only one living species; have fan-shaped leaves with branching veins and fleshy cones with seeds; ginkgoes

Division **Cycadophyta** palmlike plants; have large, featherlike leaves; produces seeds in cones; cycads

Division **Coniferophyta** deciduous or evergreen; trees or shrubs; have needlelike or scalelike leaves; seeds produced in cones; conifers

Division **Gnetophyta** shrubs or woody vines; seeds are produced in cones; division contains only three genera; gnetum

Division **Anthophyta** dominant group of plants; flowering plants; have fruits with seeds

## Kingdom Animalia

Phylum **Porifera** aquatic organisms that lack true tissues and organs; are asymmetrical and sessile; sponges

Phylum **Cnidaria** radially symmetrical organisms; have a digestive cavity with one opening; most have tentacles armed with stinging cells; live in aquatic environments singly or in colonies; includes jellyfish, corals, hydra, and sea anemones

Phylum **Platyhelminthes** bilaterally symmetrical worms; have flattened bodies; digestive system has one opening; parasitic and free-living species; flatworms

**Division Bryophyta**
Liverwort

**Division Anthophyta**
Tomato plant

**Phylum Platyhelminthes**
Flatworm

## Phylum Chordata

**Phylum Nematoda** round, bilaterally symmetrical body; have digestive system with two openings; free-living forms and parasitic forms; roundworms

**Phylum Mollusca** soft-bodied animals, many with a hard shell and soft foot or footlike appendage; a mantle covers the soft body; aquatic and terrestrial species; includes clams, snails, squid, and octopuses

**Phylum Annelida** bilaterally symmetrical worms; have round, segmented bodies; terrestrial and aquatic species; includes earthworms, leeches, and marine polychaetes

**Phylum Arthropoda** largest animal group; have hard exoskeletons, segmented bodies, and pairs of jointed appendages; land and aquatic species; includes insects, crustaceans, and spiders

**Phylum Echinodermata** marine organisms; have spiny or leathery skin and a water-vascular system with tube feet; are radially symmetrical; includes sea stars, sand dollars, and sea urchins

**Phylum Chordata** organisms with internal skeletons and specialized body systems; most have paired appendages; all at some time have a notochord, nerve cord, gill slits, and a post-anal tail; include fish, amphibians, reptiles, birds, and mammals

SCIENCE SKILL HANDBOOK

MATH SKILL HANDBOOK

FOLDABLES HANDBOOK

REFERENCE HANDBOOK

GLOSSARY/ GLOSARIO

INDEX

# Use and Care of a Microscope

**Eyepiece** Contains magnifying lenses you look through.

**Arm** Supports the body tube.

**Low-power objective** Contains the lens with the lowest power magnification.

**Stage clips** Hold the microscope slide in place.

**Coarse adjustment** Focuses the image under low power.

**Fine adjustment** Sharpens the image under high magnification.

**Body tube** Connects the eyepiece to the revolving nosepiece.

**Revolving nosepiece** Holds and turns the objectives into viewing position.

**High-power objective** Contains the lens with the highest magnification.

**Stage** Supports the microscope slide.

**Light source** Provides light that passes upward through the diaphragm, the specimen, and the lenses.

**Base** Provides support for the microscope.

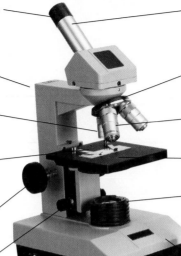

## Caring for a Microscope

1. Always carry the microscope holding the arm with one hand and supporting the base with the other hand.

2. Don't touch the lenses with your fingers.

3. The coarse adjustment knob is used only when looking through the lowest-power objective lens. The fine adjustment knob is used when the high-power objective is in place.

4. Cover the microscope when you store it.

## Using a Microscope

1. Place the microscope on a flat surface that is clear of objects. The arm should be toward you.

2. Look through the eyepiece. Adjust the diaphragm so light comes through the opening in the stage.

3. Place a slide on the stage so the specimen is in the field of view. Hold it firmly in place by using the stage clips.

4. Always focus with the coarse adjustment and the low-power objective lens first. After the object is in focus on low power, turn the nosepiece until the high-power objective is in place. Use ONLY the fine adjustment to focus with the high-power objective lens.

## Making a Wet-Mount Slide

1. Carefully place the item you want to look at in the center of a clean, glass slide. Make sure the sample is thin enough for light to pass through.

2. Use a dropper to place one or two drops of water on the sample.

3. Hold a clean coverslip by the edges and place it at one edge of the water. Slowly lower the coverslip onto the water until it lies flat.

4. If you have too much water or a lot of air bubbles, touch the edge of a paper towel to the edge of the coverslip to draw off extra water and draw out unwanted air.

SCIENCE SKILL HANDBOOK

MATH SKILL HANDBOOK

FOLDABLES HANDBOOK

REFERENCE HANDBOOK

GLOSSARY/ GLOSARIO

INDEX

# Glossary/Glosario

**Cómo usar el glosario en español:**
1. Busca el término en inglés que desees encontrar.
2. El término en español, junto con la definición, se encuentran en la columna de la derecha.

## Pronunciation Key

Use the following key to help you sound out words in the glossary.

| | | | | |
|---|---|---|---|---|
| **a** | back (BAK) | | **ew** | food (FEWD) |
| **ay** | day (DAY) | | **yoo** | pure (PYOOR) |
| **ah** | father (FAH thur) | | **yew** | few (FYEW) |
| **ow** | flower (FLOW ur) | | **uh** | comma (CAH muh) |
| **ar** | car (CAR) | | **u (+ con)** | rub (RUB) |
| **e** | less (LES) | | **sh** | shelf (SHELF) |
| **ee** | leaf (LEEF) | | **ch** | nature (NAY chur) |
| **ih** | trip (TRIHP) | | **g** | gift (GIHFT) |
| **i (i + com + e)** | idea (i DEE uh) | | **j** | gem (JEM) |
| **oh** | go (GOH) | | **ing** | sing (SING) |
| **aw** | soft (SAWFT) | | **zh** | vision (VIH zhun) |
| **or** | orbit (OR buht) | | **k** | cake (KAYK) |
| **oy** | coin (COYN) | | **s** | seed, cent (SEED, SENT) |
| **oo** | foot (FOOT) | | **z** | zone, raise (ZOHN, RAYZ) |

---

**English** Ⓐ **Español**

**absorption/asymmetry** | **absorción/asimetría**

**absorption:** the process in which nutrients from digested food are taken into the body. (p. 433)

**aggression:** a forceful behavior used to dominate or control another animal. (p. 461)

**appendage:** a structure, such as a leg or an arm, that extends from the central part of the body. (p. 387)

**asymmetry:** a body plan in which an organism cannot be divided into any two parts that are nearly mirror images. (p. 377)

**absorción:** proceso en el cual los nutrientes del alimento digerido son alojados dentro del cuerpo. (pág. 433)

**agresión:** comportamiento contundente usado para dominar o controlar otro animal. (pág. 461)

**apéndice:** estructura, como una pierna o un brazo, que se prolonga de la parte central del cuerpo. (pág. 387)

**asimetría:** plano corporal en el cual un organismo no se puede dividir en dos partes que sean casi imágenes al espejo una de otra. (pág. 377)

Science Skill Handbook

Math Skill Handbook

Reference Handbook

Glossary/ Glosario

Index

**B**

**behavior:** the way an organism reacts to other organisms or to its environment. (p. 447)

**bilateral symmetry:** a body plan in which an organism can be divided into two parts that are nearly mirror images of each other. (p. 377)

**bioluminescence (BI oh lew muh NE sunts):** the ability of certain living things to give off light. (p. 458)

**comportamiento:** forma en la que un organismo reacciona hacia otros organismos o hacia su medioambiente. (pág. 447)

**simetría bilateral:** plano corporal en el cual un organismo se puede dividir en dos partes que sean casi imágenes al espejo una de otra. (pág. 377)

**bioluminiscencia:** capacidad de ciertos seres vivos de producir luz. (pág. 458)

**C**

**chordate (KOR dat):** an animal that has a noto-chord, a nerve cord, a tail, and structures called pharyngeal pouches at some point in its life. (p. 393)

**closed circulatory system:** a system that trans-ports materials through blood using vessels. (p. 425)

**coelom (SEE lum):** a fluid-filled cavity in the body of an animal. (p. 412)

**conditioning:** a way of learning new behaviors where a behavior is modified so that a response to one stimulus becomes associated with a different stimulus. (p. 453)

**crop:** a specialized structure in the digestive system where ingested material is stored. (p. 432)

**cordado:** animal que en algún momento de su vida tiene notocordio, cordón nervioso, cola y estructuras llamadas bolsas faríngeas. (pág. 393)

**sistema circulatorio cerrado:** sistema que transporta materiales a través de la sangre usando vasos. (pág. 425)

**celoma:** cavidad llena de fluido en el cuerpo de un animal. (pág. 412)

**condicionamiento:** forma de aprender comportamientos en la cual se modifica una conducta, de tal manera que la respuesta a un estímulo se asocia con un estímulo diferente. (pág. 453)

**buche:** estructura especializada en el sistema digestivo donde el material ingerido se almacena. (pág. 432)

**D**

**diffusion:** the movement of substances from an area of higher concentration to an area of lower concentration. (p. 422)

**difusión:** movimiento de sustancias de un área de mayor concentración a un área de menor concentración. (pág. 422)

**E**

**exoskeleton:** a thick, hard outer covering; pro-tects and supports an animal's body. (p. 387)

**exoesqueleto:** cubierta externa, gruesa y dura; protege y soporta el cuerpo de un animal. (pág. 387)

SCIENCE SKILL HANDBOOK

MATH SKILL HANDBOOK

REFERENCE HANDBOOK

GLOSSARY/GLOSARIO

INDEX

**F**

**fertilization:** a reproductive process in which a sperm joins with an egg. (p. 466)

**fertilización:** proceso reproductivo en el cual un espermatozoide se une con un óvulo. (pág. 466)

**G**

**gill:** an organ that exchanges carbon dioxide for oxygen in water. (p. 423)

**gizzard:** a muscular pouch similar to a stomach that is used to grind food. (p. 432)

**branquia:** órgano que intercambia dióxido de carbono por oxígeno en el agua. (pág. 423)

**molleja:** bolsa muscular similar al estómago que sirve para triturar el alimento. (pág. 432)

**H**

**hibernation:** a response in which an animal's body temperature, activity, heart rate, and breathing rate decrease during periods of cold weather. (p. 451)

**hydrostatic skeleton:** a fluid-filled internal cavity surrounded by muscle tissue. (p. 412)

**hibernación:** respuesta en la cual la temperatura corporal, el ritmo cardíaco y la tasa de respiración de un animal disminuyen durante los periodos fríos. (pág. 451)

**hidroesqueleto:** cavidad interna llena de fluido rodeada por tejido muscular. (pág. 412)

**I**

**imprinting:** behavior that occurs when an animal forms an attachment to an organism or place within a specific time period after birth or hatching. (p. 452)

**innate behavior:** a behavior that is inherited rather than learned. (p. 449)

**instinct (IHN stingt):** a complex pattern of innate behaviors. (p. 450)

**invertebrate (ihn VUR tuh brayt):** an animal that does not have a backbone. (p. 376)

**impronta:** comportamiento que ocurre cuando un animal forma un apego a otro organismo o lugar dentro de un período específico de tiempo, después de nacer o eclosionar. (pág. 452)

**comportamiento innato:** comportamiento heredado más que aprendido. (pág. 449)

**instinto:** patrón complejo de comportamientos innatos. (pág. 450)

**invertebrado:** animal que no tiene columna vertebral. (pág. 376)

**M**

**metamorphosis (me tuh MOR fuh sihs):** a developmental process in which the body form of an animal changes as it grows from an egg to an adult. (p. 470)

**migration:** the instinctive, seasonal movement of a population of organisms from one place to another. (p. 451)

**metamorfosis:** proceso de desarrollo en el cual la forma del cuerpo de un animal cambia a medida que crece del huevo al adulto. (pág. 470)

**migración:** movimiento instintivo de temporada de una población de organismos de un lugar a otro. (pág. 451)

SCIENCE SKILL HANDBOOK

MATH SKILL HANDBOOK

REFERENCE HANDBOOK

GLOSSARY/ GLOSARIO

INDEX

**N**

**nerve net:** a netlike control system that sends signals to and from all parts of the body. (p. 414)

**notochord:** a flexible, rod-shaped structure that supports the body of a developing chordate. (p. 393)

**red nerviosa:** sistema de control parecido a una red que envía señales hacia y desde todas las partes del cuerpo. (pág. 414)

**notocordio:** estructura flexible con forma de varilla que soporta el cuerpo de un cordado en desarrollo. (pág. 393)

**O**

**open circulatory system:** a system that transports blood and other fluids into open spaces that surround organs in the body. (p. 424)

**ovary (OH va ree):** the female reproductive organ that produces egg cells; stores oocytes, which mature into ova. (p. 466)

**sistema circulatorio abierto:** sistema que transporta sangre y otros fluidos hacia espacios abiertos que rodean a los órganos en el cuerpo. (pág. 424)

**ovario:** el órgano reproductivo femenino que produce óvulos; tiendas de ovocitos que maduran en los óvulos. (pág. 466)

**P**

**pheromone (FER uh mohn):** a chemical that is produced by one animal and influences the behavior of another animal of the same species. (p. 459)

**feronoma:** químico que es producido por un animal y que influye en el comportamiento de otro animal de la misma especie. (pág. 459)

**R**

**radial symmetry:** a body plan in which an organism can be divided into two parts that are nearly mirror images of each other anywhere through its central axis. (p. 377)

**simetría radial:** plano corporal en el cual un organismo se puede dividir en dos partes para que sean casi imágenes al espejo una de la otra, en cualquier parte del eje axial. (pág. 377)

**S**

**sexual reproduction:** type of reproduction in which the genetic material from two different cells—a sperm and an egg—combine, producing an offspring. (p. 465)

**society:** a group of animals of the same species living and working together in an organized way. (p. 460)

**spiracle:** a tiny hole on the surface of an organism where oxygen enters the body and carbon dioxide leaves the body. (p. 422)

**reproducción sexual:** tipo de reproducción donde el material genético de dos células diferentes— un espermatozoide y un óvulo—se combinan para dar origen a una cría. (pág. 465)

**sociedad:** grupo de animales de la misma especie que viven y trabajan juntos de una forma organizada. (pág. 460)

**espiráculo:** hueco diminuto en la superficie de un organismo por donde entra oxígeno al cuerpo y sale dióxido de carbono. (pág. 422)

SCIENCE SKILL HANDBOOK

MATH SKILL HANDBOOK

REFERENCE HANDBOOK

GLOSSARY/ GLOSARIO

INDEX

**T**

**territory:** an area that is set up and defended by animals for feeding, mating, and raising young. (p. 461)

**testis:** the male reproductive organ that produces sperm. (p. 466)

**territorio:** área que un grupo de animales establece y defiende para alimentarse, aparearse y criar su descendencia. (pág. 461)

**testículo:** el órgano reproductor masculino que produce espermatozoides. (pág. 466)

**U**

**undulation (un juh LAY shun):** the wavelike motion of some animals. (p. 416)

**ondulación:** movimiento de algunos animales parecido a una ola. (pág. 416)

**V**

**vertebrate (VUR tuh brayt):** an animal with a backbone. (p. 376)

**vertebrado:** animal con columna vertebral. (pág. 376)

**Z**

**zygote:** the new cell that forms when a sperm cell fertilizes an egg cell. (p. 466)

**zigoto:** célula nueva que se forma cuando un espermatozoide fertiliza un óvulo. (pág. 466)

SCIENCE SKILL HANDBOOK

MATH SKILL HANDBOOK

REFERENCE HANDBOOK

GLOSSARY/ GLOSARIO

INDEX

# Index

*Italic numbers* = illustration/photo    **Bold numbers** = vocabulary term
*lab* = indicates entry is used in a lab on this page

## A

**Absorption, 433**
**Academic Vocabulary,** 383, 424, 465
**Aggression,** *461,* **461**
**Amphibia,** 395
**Amphibian(s)**
    explanation of, **395,** *395*
    heart in, 425
**Animal(s)**
    characteristics of, 375, 375 *lab* , 411, 427
    circulation in, 421, *424,* 424–425, *425*
    classification of, *376,* 376–378, 376 *lab, 377, 378, 379*
    control structures in, *414,* 414–415, *415*
    designed for unique environments, 436–437 *lab*
    gas exchange in, 421–423, *423*
    reproduction in, 375
    symmetry in, 377, *377*
    types of movement in, *416,* 416–417, *417,* 419
**Animal behavior(s)**
    based on seasons, 451
    conditions that change, 472–473 *lab*
    courtship as, 461
    explanation of, **447**
    innate, 449–451, *450, 451*
    learned, 452–453, *453*
    societies and, *460,* 460–461, *461*
    stimuli and responses and, *448,* 448–449, *449*
    territorial, 461
**Annelid worm(s),** *386, 386*
**Annelida,** 386
**Appendages**
    explanation of, **387**
    jointed, 387, 388 *lab*
**Arachnid(s),** 388
**Arthropod(s),** *387,* 387–388, *388*
**Asexual reproduction,** 465
**Asymmetry,** *377,* **377**
**Aves,** 397
**Backbones.** *See also* **Vertebrates**
    explanation of, 412
    in vertebrates, 393 *lab*

## B

**Baleen,** 431
**Bat(s),** 381
**Beak(s),** 397
**Behaviors.** *See* **Animal behaviors**
**Big Idea,** 372, 402
    Review, 405

## Bilateral symmetry

    in arthropods, 387
    explanation of, *377,* **377**
    in flatworms and roundworms, 385
    in mollusks and annelids, 386
**Bioluminescence,** **458**
**Bird(s)**
    courtship displays in, 461, 463, *463*
    explanation of, 397, *397*
**Body language,** 459
**Bony fish,** 394
**Brain(s)**
    animals with, 415, *415*

## C

**Careers in Science,** 381
**Cartilage,** 394, 413
**Cartilaginous fish,** 394
**Cell division,** 468
**Cephalopod(s),** 419
**Change**
    responses to, 448, *448*
**Chapter Review,** 404–405, 440–441, 476–477
**Chemical(s)**
    animal production of, 459
**Chordata,** 378
**Chordate(s)**
    characteristics of, 393
    explanation of, **393**
**Circulation,** 421, 424
**Circulatory system(s)**
    closed, *424,* 425
    explanation of, 424
    open, 424, *424*
**Closed circulatory system**
    absorption in, 433
    explanation of, *424,* 425
**Cnidarian(s) ,** 384, *384,* 414
**Coelom,** 412
**Cognitive behavior,** 453
**Common Use.** *See* **Science Use v. Common Use**
**Communication**
    methods of animal, *457,* 457–459, *458, 459*
**Conditioning,** **453**
**Control**
    structures in animals for, *414,* 414–415, *415*
**Courtship**
    displays of, 463, *463*
    explanation of, *461,* **461**
**Crop,** **432**
**Crustacean(s)**
    explanation of, 388, *388*
    use of bioluminescence by, 458
**Cycle,** 465

## D

**Development**
    explanation of, 468
    gestation as, 469, *469*
    internal, 468, *468,* 469, *469*
    metamorphosis as, 470
**Dichotomous key,** 391
**Diffusion**
    explanation of, **422**
    release of carbon dioxide by, 434
**Digestion**
    absorption and, 433
    function of, 429
    structures for, *430,* 430–432, *431*
**Dominance,** **460**

## E

**Earthworm(s)**
    change in behavior of, 472–473 *lab*
    hydrostatic skeletons in, 412, *412*
    movement of, 411 *lab*
**Echinoderm(s),** 389, *389*
**Echinodermata,** 389
**Egg(s)**
    explanation of, **397,** 465
    fertilization of, 466
**Embryo(s)**
    development of, 469
    explanation of, 468, *468*
**Endoskeleton(s),** 413, *413*
**Enzyme(s),** 433
**Exchanged,** 424
**Excretion**
    in aquatic animals, 434
    function of, 429, 434, *434*
    in terrestrial animals, 434
**Exoskeleton(s),** **387,** 413, *413*
**External fertilization,** 467, *467*

## F

**Feeding**
    structures for, *430,* 430–431, *431*
**Female(s)**
    reproductive organs in, 466
**Fertilization**
    explanation of, 466
    external, 467
    internal, 467
**Filter feeding,** 431
**Fins(s),** 394
**Fish**
    use of bioluminescence by, 458
    characteristics of, 394, *394*
    filter feeding in, 431

# Credits

## Photo Credits

**Front Cover** Spine Photodisc/Getty Images; **Back Cover** ThinkStock/Getty Images; **Inside front,back cover** ThinkStock/Getty Images; **Connect Ed** (t) Richard Hutchings, (c)Getty Images, (b)Jupiterimages/ThinkStock/Alamy; **i** ThinkStock/Getty Images; viii–ix The McGraw-Hill Companies; **ix** (b)Fancy Photography/Veer; **370** Wayne R Bilenduke/Getty Images; **372–373** Gary Bell/Getty Images; **373** Darrell Gulin/Getty Images; **375** (t)Hutchings Photography/Digital Light Source, (b)John Cancalosi/Alamy; **376** Geoff Brightling/Getty Images; **377** (l)Brand X Pictures/PunchStock, (c)Paul Nicklen/National Geographic/Getty Images, (r)Design Pics/Carson Ganci; **378** (t)Ingram Publishing/Alamy, (c)Francesco Rovero, (b)Frank Greenaway/Getty Images; **380** (t)John Cancalosi/Alamy, (b)Paul Nicklen/National Geographic/Getty Images; **381** (c)age fotostock/SuperStock, (inset)N. Simmons/American Museum of Natural History, (bkgd)Heath Korvola/Getty Images; **382** Matthew Oldfield, Scubazoo/Photo Researchers; **383** Hutchings Photography/Digital Light Source; **385** (t)James H. Robinson/Photo Researchers, (b)Sinclair Stammers/Photo Researchers; **386** (t)Ingram Publishing/Alamy, (b)Doug Perrine/SeaPics.com; **388** (t)Ingram Publishing/Alamy, (c)Hutchings Photography/Digital Light Source, (b)Jeremy Woodhouse/Getty Images; **390** (t)Ingram Publishing/Alamy, (b)Jeremy Woodhouse/Getty Images; **391** (l to r, t to b)C Squared Studios/Getty Images, (1)Siede Preis/Getty Images, (2)Brand X Pictures/PunchStock, (3)John Foxx/Alamy, (4)Photodisc/Getty Images, (5)Creatas/PunchStock, (6)Hutchings Photography/Digital Light Source, (7)Siede Preis/Getty Images, (8)Brand X Pictures/PunchStock, (9)Siede Preis/Getty Images; **392** FRANS LANTING/National Geographic Stock; **393** Hutchings Photography/Digital Light Source; **394** Pacific Stock/SuperStock; **396** (inset)Daniel Heuclin/NHPA/Photoshot, (bkgd)JH Pete Carmichael/Getty Images; **398** Steve Kaufman/CORBIS; **399** (t)Daniel Heuclin/NHPA/Photoshot, (c)Steve Kaufman/CORBIS, (b)www.flickr.com/photos/twobigpaws/Flickr/Getty Images; **400** (t,c)Macmillan/McGraw-Hill, (b) Hutchings Photography/Digital Light Source; **401** Hutchings Photography/Digital Light Source; **402** (t)John Cancalosi/Alamy, (b)Doug Perrine/SeaPics.com; **404** (l)Frederic Pacorel/Getty Images, (r)IT Stock Free/Alamy; **405** (l)Ingram Publishing/Alamy, (c)David Fleetham/Visuals Unlimited/Getty Images, (r)Gary Bell/Getty Images; **408–409** J. Keith Rankin/Alamy; **410** Neil Lucas/Minden Pictures; **411** (t)Mark Steinmetz, (b)Vincent Leblic/Photolibrary; **413** (t)Jeff Foott/Getty Images, (b)Dorling Kindersley/Getty Images; **414** Hutchings Photography/Digital Light Source; **416** (l)Mark Strickland/SeaPics.com, (r)Ken Usami/Getty Images; **417** (t)Art Wolfe/Getty Images, (c)Nicholas Bergkessel, Jr./Photo Researchers, (b)Brand X Pictures/PunchStock; **418** (t)Vincent Leblic/Photolibrary, (b)Brand X Pictures/PunchStock; **419** Prisma/SuperStock **420** Microfield Scientific Ltd/Photo Researchers; **421** Hutchings Photography/Digital Light Source; **422** Hutchings Photography/Digital Light Source; **427** Macmillan/McGraw-Hill; **428** Martin Shields/Alamy; **429** (t)Ken Lucas/Visuals Unlimited, (b) Lester Lefkowitz/Getty Images; **430** (t)Heinrich van den Berg/Getty Images, (c)E. & P. Bauer/CORBIS, (b)Merlin Tuttle/BCI/Photo Researchers; **431** (l)Bartomeu Borrell/age fotostock, (r)H. Robinson James/Getty Images; **432** The Mcgraw-Hill Companies; **434** Creatas/PunchStock; **435** (t)Merlin Tuttle/BCI/Photo Researchers, (b)Creatas/PunchStock; **436** The McGraw-Hill Companies; **437** Hutchings Photography/Digital Light Source; **438** Dorling Kindersley/Getty Images; **440** (l)John Foxx/Alamy, (r)Creatas/PunchStock; **441** Scott Linstead/Tom Stack & Associates; **444–445** Anup Shah/Minden Pictures; **446** Worldwide Picture Library/Alamy; **447** (t)Hutchings Photography/Digital Light Source, (b)Eric Gay/AP Images; **448** (l)Mark Steinmetz, (r)Joseph Devenney/Getty Images; **449** Steve Bloom/Getty Images; **450** Bianca Lavies/National Geographic/Getty Images; **451** age fotostock/SuperStock; **453** (l,r)Mark Steinmetz, (c)Thomas & Pat Leeson/Photo Researchers; **454** (t)Joseph Devenney/Getty Images, (c)Bianca Lavies/National Geographic/Getty Images, (b)Mark Steinmetz; **455** (t,tc) Macmillan/McGraw-Hill, (b,bc)Hutchings Photography/Digital Light Source; **456** Anna Henly/Getty Images; **457** (t)Hutchings Photography/Digital Light Source, (b)Photography EPC/Photolibrary; **458** (t)Hutchings Photography/Digital Light Source, (b)Satoshi Kuribayashi/Photolibrary; **460** SUZI ESZTERHAS/MINDEN PICTURES/National Geographic Stock; **461** (t)Arco Images GmbH/Alamy, (b)Dave Watts/Tom Stack & Associates; **462** (t)SUZI ESZTERHAS/MINDEN PICTURES/National Geographic Stock, (c)Dave Watts/Tom Stack & Associates, (b)Arco Images GmbH/Alamy; **463** (inset)TIM LAMAN/National Geographic Stock, (bkgd)Getty Images; **464** Thomas Marent/Minden Pictures; **465** (l to r, t to b)David J Green/Getty Images, (1)Purestock/PunchStock, (2)NPS Photo, (3)IT Stock/age fotostock, (4)Richard Wear/Design Pics/CORBIS, (5)CORBIS, (6)Leonard Lee Rue III/Photo Researchers; **466** (l)Tim Hawley/Getty Images, (r)Michael Winokur/Getty Images; **467** John Cancalosi/Photolibrary; **468** (inset)Daniel Heuclin/NHPA/Photoshot, (bkgd)PHONE PHONE-Auteurs Cordier Sylvain/Peter Arnold, Inc.; **469** David Higgs/NHPA/Photoshot; **471** (t)Leonard Lee Rue III/Photo Researchers, (tc)John Cancalosi/Photolibrary, (b)John Cancalosi/Photolibrary, (bc)PHONE PHONE-Auteurs Cordier Sylvain/Peter Arnold, Inc.; **472** (l to r, t to b,2,5,6)Macmillan/McGraw-Hill, (others)Hutchings Photography/Digital Light Source; **473** Hutchings Photography/Digital Light Source; **474** (t)Thomas & Pat Leeson/Photo Researchers, (c)Anna Henly/Getty Images, (b)Daniel Heuclin/NHPA/Photoshot; **476** Bianca Lavies/National Geographic/Getty Images; **477** (l)Dave Watts/Tom Stack & Associates, (r)Anup Shah/Minden Pictures; **SR-00–SR-01** (bkgd)Gallo Images-Neil Overy/Getty Images; **SR-02** Hutchings Photography/Digital Light Source; **SR-06** Michell D. Bridwell/PhotoEdit; **SR-07** (t)The McGraw-Hill Companies, (b)Dominic Oldershaw; **SR-08** StudiOhio; **SR-09** Timothy Fuller; **SR-10** Aaron Haupt; **SR-42** (c)NIBSC / Photo Researchers, Inc., (r) Science VU/Drs. D.T. John & T.B. Cole/Visuals Unlimited, Inc.; Stephen Durr; **SR-43** (t)Mark Steinmetz, (r)Andrew Syred/Science Photo Library/Photo Researchers, (br)Rich Brommer; **SR-44** (l)Lynn Keddie/Photolibrary, (tr)G.R. Roberts; David Fleetham/Visuals Unlimited/Getty Images; **SR-45** Gallo Images/CORBIS; **SR-46** Matt Meadows.